THE TROPICAL AGRICU

General Editor, Livestock Vol

Anthony J. Smith

Pigs

Revised Edition

David H. Holness

Livestock Consultant, Harare, Zimbabwe

Macmillan Education
Between Towns Road, Oxford OX4 3PP
A division of Macmillan Publishers Limited
Companies and representatives throughout the world

www.macmillan-africa.com
www.macmillan-caribbean.com

Published in co-operation with the ACP–EU Technical Centre for Agricultural and Rural Cooperation (CTA), Postbus 380, 6700 AJ Wageningen, The Netherlands

ISBN 978-0-333-79148-6

First published 1991
Reprinted 1993
Second (revised) edition 2005

Editing and layout: Green Ink Ltd, UK

Cover design: Jim Weaver Design

Cover photograph: Landrace piglets, courtesy of NHPA

All photographs are reproduced by courtesy of the author except the following: Figs 13, 14, 17, 18, 21 and 59 by A. Smith; Figs 52 and 54 by J. Thomson.

Printed and bound in Malaysia

2013 2012 2011
10 9 8 7 6 5 4 3

Technical Centre for Agricultural and Rural Cooperation (ACP-EU)

The Technical Centre for Agricultural and Rural Cooperation (CTA) was established in 1983 under the Lomé Convention between the ACP (African, Caribbean and Pacific) Group of States and the European Union Member States. Since 2000, it has operated within the framework of the ACP–EC Cotonou Agreement.

CTA's tasks are to develop and provide services that improve access to information for agricultural and rural development, and to strengthen the capacity of ACP countries to produce, acquire, exchange and utilise information in this area. CTA's programmes are designed to: provide a wide range of information products and services and enhance awareness of relevant information sources; promote the integrated use of appropriate communication channels and intensify contacts and information exchange (particularly intra-ACP); and develop ACP capacity to generate and manage agricultural information and to formulate information and communication management strategies, including those relevant to science and technology. CTA's work incorporates new developments in methodologies and cross-cutting issues such as gender and social capital.

CTA, Postbus 380, 6700 AJ Wageningen, The Netherlands.

Titles in *The Tropical Agriculturalist* series

Title	ISBN
Animal Breeding	ISBN 0-333-57298-X
Animal Health Vol.1	0-333-61202-7
Animal Health Vol.2	0-333-57360-9
Camels	0-333-60083-5
Dairying	0-333-52313-X
Donkeys	0-333-62750-4
Draught Animals	0-333-52307-5
Goats	0-333-52309-1
Livestock Behaviour, Management and Welfare	0-333-62749-0
Livestock Production Systems	0-333-60012-6
Pigs Revised Edition	0-333-79148-7
Poultry Revised Edition	0-333-79149-5
Rabbits	0-333-52311-3
Ruminant Nutrition	0-333-57073-1
Sheep Revised Edition	0-333-79881-3
Tilapia	0-333-57472-9
Warm-water Crustaceans	0-333-57462-1

Title	ISBN
Alley Farming	0-333-60080-0
Avocado	0-333-57468-0
Cassava	0-333-47395-7
Chickpeas	0-333-63137-1
Cocoa	0-333-57076-6
Coconut	0-333-57466-4
Coffee Growing	0-333-54451-X
Cotton	0-333-47280-2
Cut Flowers	0-333-62528-5
Food Crops and Drought	0-333-59831-8
Food Legumes	0-333-53850-1
Forage Husbandry	0-333-66856-1
Groundnut	0-333-72365-1
Maize	0-333-44404-3
Market Gardening	0-333-65149-8
Oil Palm	0-333-57465-6
Plantain Bananas	0-333-44813-8
Rubber	0-333-68355-2
Sorghum	0-333-54452-8
Spice Plants	0-333-57460-5
Sugar Cane	0-333-57075-8
Sweet Potato	0-333-79150-9
Tea	0-333-54450-1
The Storage of Food Grains and Seeds	0-333-44827-8
Upland Rice	0-333-44889-8
Weed Control	0-333-54449-8

Other titles published by Macmillan with CTA

Title	ISBN
Animal Production in the Tropics and Subtropics	ISBN 0-333-53818-8
Coffee: The Plant and the Product	0-333-57296-3
Controlling Crop Pests and Diseases	0-333-57216-5
Dryland Farming in Africa	0-333-47654-9
The Tropical Vegetable Garden	0-333-57077-4
The Yam	0-333-57456-7
Where There is No Vet	0-333-58899-1

Contents

Preface

Since the first edition of *Pigs* was published in 1991, rapid improvements in communication have taken place all around the world. It is now much easier to transfer information from developed to developing regions and this includes information on pig production. Some technological developments offer exciting prospects for improving pig production in the tropics, while others are totally inappropriate, particularly for small-scale producers. In this book, the author has identified which developments could be of value to tropical pig producers and which should be avoided.

Some new practices in developing countries have proved to be undesirable. For example, the identification of beef from Bovine Spongiform Encephalopathy (BSE) infected cattle as the probable source of new variant Creutzfeldt-Jakob disease (CJD) in humans has called into question the practice of feeding animal products to other animals, including pigs. Concepts of animal welfare in parts of Europe that have led to changes in pig production systems, such as the banning of sow stalls and antibiotic growth promoters, may have an impact on methods of keeping pigs in less developed countries, particularly if these countries market their products in the North.

With these and other considerations in mind, the second edition of *Pigs* has been updated throughout and new sections added. For example, Chapter 2 includes new work on the effect of the immune system on pig productivity. New genetic improvement techniques and changes in selection priorities appear in Chapter 4. In Chapter 5, the section on trace minerals has been expanded and incorporates chelated trace minerals and their effects on mineral availability and uptake. A new section deals with feed additives and their influence on productivity and health, while another discusses genetically modified feeds.

Chapter 6 expands on specialist weaner accommodation and addresses pig welfare considerations in relation to housing design. Changes in approaches to some aspects of disease control in the tropics are dealt within Chapter 7. Finally, a new section in Chapter 9 looks at new marketing opportunities that could be exploited in the tropical situation.

This thoroughly updated edition is very readable and informative and will be valued by traditional pig producers and smallholders or businessmen planning more intensive systems of production. Development workers, non-government organisations (NGOs) and extension services will also find this book useful.

Anthony Smith, September 2004

Acknowledgements

This book owes much to the experience I gained over many years of visiting and discussing problems with pig producers, both large and small, throughout Zimbabwe and surrounding territories. The period I served as Director of the Pig Industry Board of Zimbabwe was particularly fruitful in this respect.

I would like to acknowledge contributions of information and ideas from Dr Noël Chabeuf of Institut d'Elevage et de Médecine Vétérinaire des Pays Tropicaux, France, Dr Raul H. Godoy Montañez of the Unversidad Autónoma de Yucatán in Mexico, Dr Nimal Pathiraja of the University of Zimbabwe, Mr Rex Parry of Macmillan Education, Dr Tony Smith, formerly of the Centre for Tropical Veterinary Medicine in Edinburgh, Scotland, and the staff of the International Livestock Centre for Africa in Ethiopia (now known as the International Livestock Research Institute). Other technical contributions have been acknowledged in the text.

I am particularly indebted to Ms Lyn Purves who read the manuscript and made many helpful suggestions and corrections. My grateful thanks also go to Ms Grace Kahari and Ann Smythe for their typing and help with the preparation of the manuscript, and to Ms Janine Holness for the original illustrations.

The author and publishers would like to thank Dr S.H.B. Lebbie of the International Livestock Centre for Africa (ILCA), Addis Ababa for advice and comments on the manuscript, and Dr R.T. Wilson, also of ILCA, for arranging Dr Lebbie's review. Thanks are also due to the author and publishers of Whittemore, C.T. (1987) *Elements of Pig Science* (Longman) for permission to redraw their Fig 2.16.

1 Distribution, potentials and constraints

World pig population

The estimated pig population of 913 million (FAO, 1999) means that there is approximately one pig for every six people in the world. Although pigs are numerically fewer than some other domestic species, more pig meat is produced then any other meat (Table 1). This reflects the greater productivity of the pig when compared with other domestic species.

Table 1 Numbers and meat production of the main domestic livestock species

	Numbers (millions)	Meat output ('000 metric tonnes per year)
Pigs	913	88430
Poultry	14139	63249
Cattle	1338	55867
Sheep	1069	7474
Goats	709	3821
Buffalo	158	3083

Source: FAO (1999)

Distribution and consumption

The distribution of pigs in the world is not uniform. Some 58 per cent of the world's pig population is found in the Asia–Pacific region, with a further 22 per cent in Europe and countries of the former Soviet Union. In contrast, the number of pigs in large parts of the tropical and subtropical developing regions (e.g. Africa and Latin America) is relatively small (Fig 1). However, during the last decade, the increase in global pig numbers has been largely attributable to growth in the developing world,

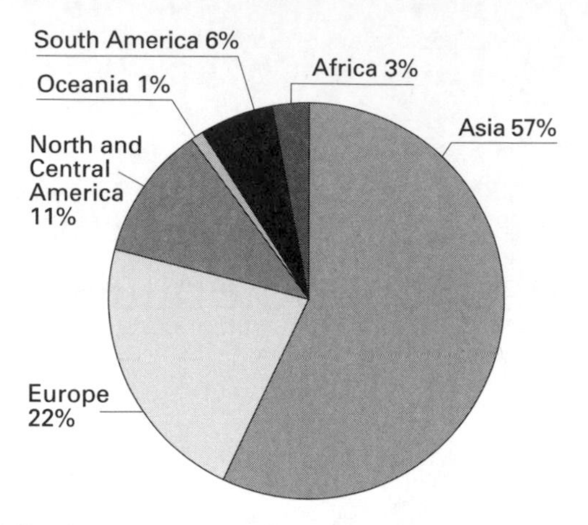

Fig 1 *Regional distribution of the world's pig population (FAO, 1999)*

which now has some 65 per cent of the world's population of pigs (Fig 2). Growth in numbers has slowed in the developed countries, mainly due to concerns over pollution and environmental awareness. China remains the predominant pig producing country, but other areas, such as Brazil, Argentina and India, are emerging rapidly. Brazil, for example, now has the same number of pigs per inhabitant (0.22) as the USA.

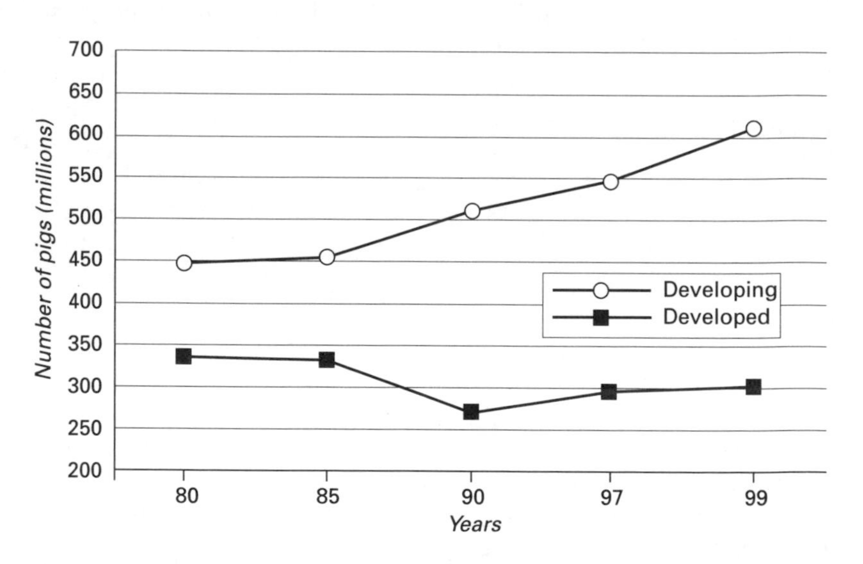

Fig 2 *Changes in the pig population in developed and developing regions of the world 1980–1999 (FAO, 1999)*

Similarly, there are marked differences in patterns of consumption of pig meat in different regions of the world. In some parts of Europe, annual per capita consumption of pig meat is over 50 kg, and represents some 60 per cent of the total meat consumed. At the other end of the scale, in areas of the developing world (particularly in Africa), estimated annual per capita consumption ranges from 2 to 3 kg, and forms less than 10 per cent of the total meat diet.

The reasons for the uneven distribution of pigs throughout the tropical and sub-tropical world are manifold. In tropical Asia and parts of China, pork is the main animal protein component of the diet. On the other hand, in areas where the Islamic religion prevails, such as the Middle East, Pakistan and parts of Africa, Muslims are forbidden to eat pig meat. Similarly, believers in the Jewish faith are instructed not to eat pork. Social factors also play a part and these may have a positive or a negative effect on the pig population. In some Pacific islands, such as Tonga and Papua New Guinea, pigs are highly regarded as a source of wealth and are associated with marriage customs. In Africa, people have traditionally obtained meat mainly from ruminants, particularly cattle, and this preference persists.

The pig has historically been considered an unclean animal, wallowing in filth, an object of distaste and a hazard to human health. Clearly there is some truth in this assumption if the pig is kept as a scavenger, but the exact opposite pertains if the pig is well managed under confined conditions.

Climate also has an influence on distribution. With suitable housing and management, pigs can be reared almost anywhere, but in situations of extreme temperatures, high humidity or lack of rainfall, production costs will rise because of the need for housing and lack of availability of suitable feeds.

Potentials and constraints in developing countries

The world trend is towards the consumption of more white than red meat. Thus the potential for increasing meat production from pigs in the developing world is enormous. When compared with cattle and other ruminants, pigs have several major advantages:

- They produce meat without contributing to the deterioration of natural grazing lands. This is of paramount importance in relation to the current levels of desertification, soil erosion and loss of productive land in tropical and sub-tropical parts of the world. Overstocking and consequent overgrazing by ruminants is a primary cause of this degradation.

- They convert concentrated feed to meat twice as efficiently as ruminants.
- They have the potential to be highly productive. They are capable of producing large litters after a relatively short gestation period, have a short generation interval and grow rapidly, so their output in terms of yield of meat per tonne live weight of breeding females per year is in the region of six times that of cattle.
- If confined, maximum use can be made of their manure and effluent (see Chapter 3).
- Their relatively small size, when compared with cattle, provides for more flexibility in marketing and consumption.
- Pig meat is particularly suitable for processing. Many processed products have a longer shelf life than fresh meat and can thus be distributed to a wider section of the population.
- Pigs offer a quicker turnover rate on investment when compared with cattle.

Apart from the social and religious constraints already mentioned, other constraints to pig production include:

- As simple-stomached animals, they compete directly with humans for food, especially the staple grains and oilseeds. This constraint can be partly overcome by making maximum use of crop by-products, waste food and grain that is unsuitable for human consumption.
- They cannot provide a source of draught power for farming operations.
- Since they tend to be raised close to human habitation, their effluent may cause a pollution problem.
- Pigs and man are co-hosts to a number of parasites. If pigs are not confined, they can pose a danger to human health.

Economic and planning considerations

Clearly, the potential profitability of pig production will vary considerably from country to country and between areas within a country. Nevertheless, whether a prospective pig producer in the tropics is planning a small or large enterprise, he should first ask himself a few basic questions. Only when these questions have been satisfactorily answered should a start be made in pig production.

Here are some pertinent questions:

- Have I got an assured market for my pigs and is the price adequate? Can I get my pigs to market without incurring too high a cost or, alternatively, can I site the enterprise closer to the market? Can the market be expanded in the future to provide an increased income?

- Have I sufficient funds to get started? (Money is required for fixed capital costs, i.e. buildings and equipment and working capital to pay for stock, feed and labour, for the first year of operation. For a breeding/feeding unit, cash outflow tends to increase to a maximum at the end of the first 12 months of operation, while cash inflow begins only 12 months after commencement of the enterprise.)
- Can I integrate the pig unit with any other form of production in order to add value to the total enterprise?
- Can I obtain suitable stock for the type of system envisaged?
- Will these supplies be reliable and available at a suitable cost?
- Have I got access to adequate and suitable feed resources?
- Are water supplies adequate in quality and quantity?
- Do I enjoy working with animals, particularly pigs? (A requirement for good stockmanship.)
- Have I sufficient management ability? (Success in the pig enterprise depends on attention to detail, while the labour force in the tropics is often unskilled and untrained.)

2 Understanding the pig

It is essential for anybody involved in the breeding and rearing of pigs to have a basic understanding of the most important aspects of the biology of the animal. Only then will the prospective pig keeper have an appreciation of the likely husbandry requirements of the pig.

Origins and evolution

Although its exact origins are obscure, it is probable that the domestic pig is descended from the European wild boar (*Sus scrofa*). Pigs originally colonised areas of forest and swamp (as does *Sus scrofa* to this day), and thus evolved to live in a moist, shady environment. Their short legs and powerful streamlined body are built for moving through dense undergrowth, and their strong head and tusks, with a cartilaginous disc in the snout, are well suited to digging and rooting.

Biblical writings indicate that pigs were first domesticated as early as 2000 BC. As man has domesticated the pig for meat, major changes in conformation have occurred from the typical 'unimproved' type. The relatively large, narrow head, heavy forequarters, tapering light hindquarters and compact body have been replaced by a smaller head, lighter forequarters, a longer and wider body, and well developed, meaty hindquarters (Fig 3).

Feeding and digestion

Although pigs kept in tropical regions may eat a lot of fibre, stockowners should remember that pigs are simple-stomached animals. Unlike ruminants, they do not have a complex stomach with a large microbial population, which enables large quantities of fibrous material to be digested (Fig 4). The ability of pigs to digest and utilise fibre is restricted by the microbial population in the caecum, which is of relatively small volume

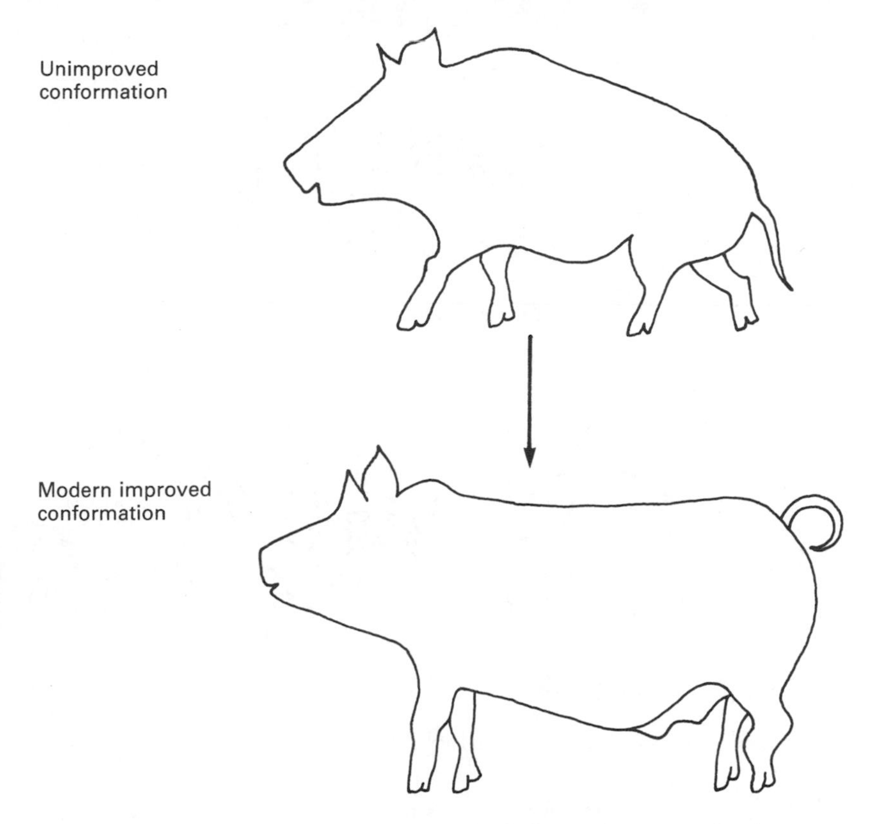

Fig 3 *Change in body conformation as a result of selection for meat production*

when compared with the rumen. It has been claimed that unimproved breeds found in Africa have an enhanced ability to utilise fibrous feeds compared with exotic breeds. While this may be so to a small extent, there are no anatomical differences in the digestive tract between the two types. Accordingly, with all pigs, high-fibre diets will have the effect of diluting the amount of nutrients available to the animal. In contrast to ruminants, pigs are unable to utilise non-protein sources of nitrogen for the production of microbial protein in the rumen. This makes them dependent on both the amount and quality of protein in their diet.

The alimentary canal, or digestive tract, of the pig is designed to digest and absorb concentrated feed. Feed taken in at the mouth is ground into a pulp by mastication. At the same time, it is moistened and mixed with saliva. Saliva contains the enzyme ptyalin, which initiates the breakdown of starch to simpler carbohydrates. The feed then passes into the stomach, which provides an acid environment. Gastric juice contains the enzyme pepsin, which begins the breakdown of protein.

PIG

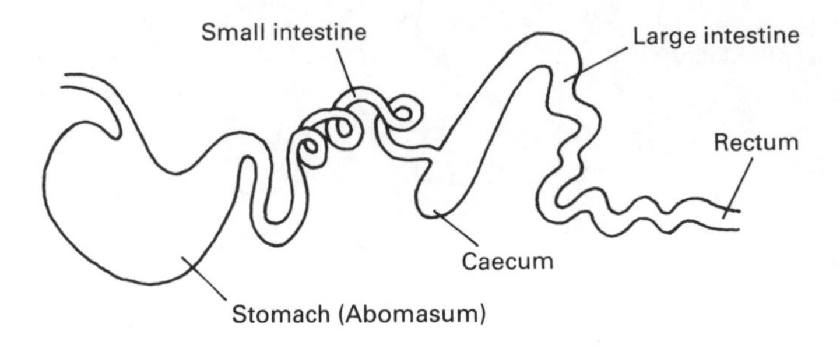

RUMINANT

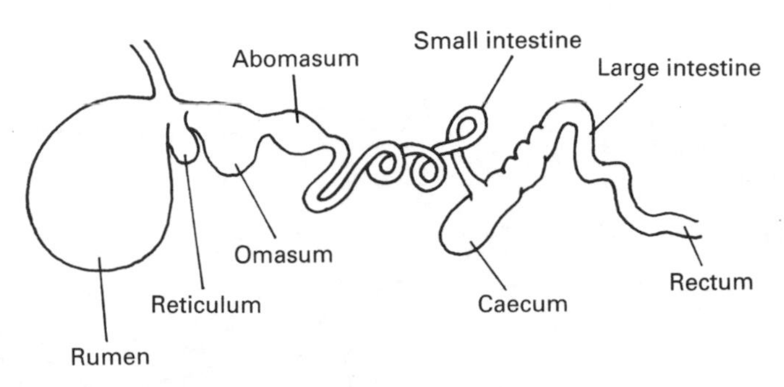

Fig 4 *Comparison of the digestive tracts of pigs and ruminants*

The small intestine is the major site where feed absorption occurs. Digestive juices from the pancreas, liver and small intestine complete the process of digestion as follows:

- Starch is hydrolysed to maltose by amylase from the pancreatic juice. Maltose and other disaccharide sugars are broken down by specific enzymes in the intestinal juice, e.g. maltase, lactase and sucrase, into monosaccharides such as glucose and fructose. These are then absorbed through the gut wall.
- Trypsin in the pancreatic juice acts on protein to produce polypeptides, which are then broken down to amino acids by various peptidases in the intestinal juice and subsequently absorbed.
- Bile, which is secreted by the liver, serves to emulsify fats into smaller globules, which are then broken down by the enzyme lipase into fatty acids and glycerol ready for absorption. Lipase is present in both the pancreatic and intestinal juices.

Pigs are omnivores and will consume a wide range of feeds from both plant and animal sources. The natural inclination of the pig is to eat on a 'little and often' basis, and this is likely to maximise both total feed intake and the efficiency of feed utilisation.

In practical terms, growth is measured as the increase in bodyweight over time, and is largely dependent on the amount of feed or total nutrient intake. However, there are major differences between the feed intake of different breeds of pig and this affects their growth response per unit of feed ingested. Because man has selected pigs for high growth rates in order to improve biological efficiency, he has selected for a larger mature size. In consequence, unimproved types of pig (common in developing parts of the world), which have not been selected for increased growth rates, will tend to grow more slowly to a smaller mature size (A in Fig 5) when compared with improved breeds (B in Fig 5). It follows that if unimproved pigs are slaughtered at the same weight as their exotic counterparts, they will be relatively more mature and therefore at a different stage of development.

The way in which the pig develops is just as important as the rate of growth. Selection has resulted in a greater propensity to lay down protein tissue in improved breeds. Thus the plateau for maximum growth potential in an improved breed can be 600 g per day compared with 400 g per day for an unimproved breed (Fig 6). As the level of feed intake increases, the unimproved pig will deposit more fat (A in Fig 6), in comparison with improved types (B in Fig 6). Because too much fat is undesirable for consumers and costly to produce (approximately five times the nutrient cost of lean tissue deposition), it is critical that pigs are fed according to their ability to grow and lay down lean tissue.

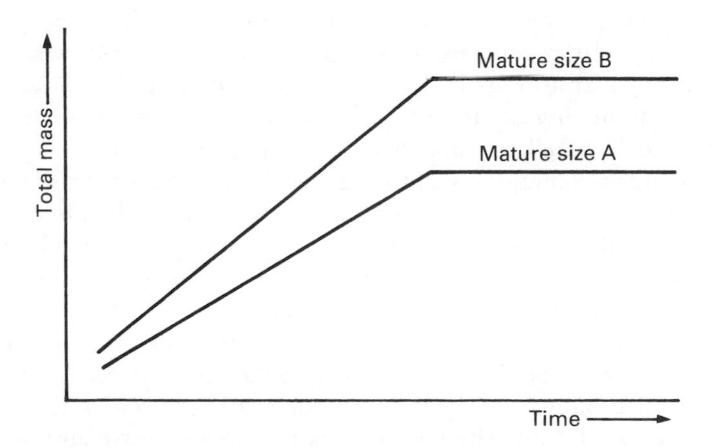

Fig 5 *Differences in mature size and consequences for growth rate in (A) unimproved and (B) improved breeds of pig*

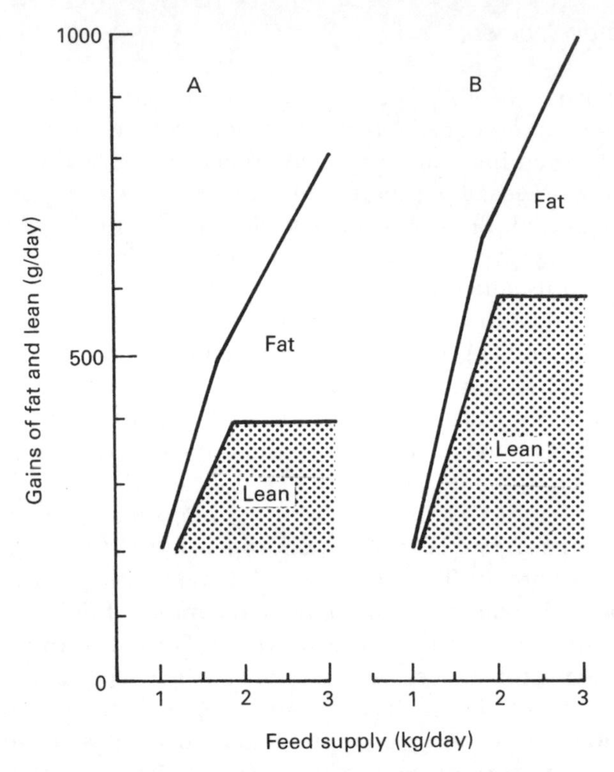

Fig 6 *Typical differences between (A) unimproved and (B) improved pig breeds in their ability to lay down lean and fat tissue as feed intake is increased (After Whittemore, 1987)*

Entire male pigs grow faster, have leaner carcasses and convert feed more efficiently than females. However, if males are castrated, the opposite is true. Traditionally, boars have been castrated in order to improve carcass quality and to prevent 'boar taint' or odour in the meat, which tends to occur as boars approach puberty. Nowadays, modern pigs grow faster and are slaughtered at younger ages and the problem of taint is considerably reduced. Unless pigs are grown slowly (as is the case when they are kept under scavenging conditions) or are required for a highly sophisticated market, there would appear to be no justification for castration in pigs destined for meat production.

Pigs are born with less than 2 per cent fat in their bodies, which makes them particularly susceptible to cold stress. As they grow, they deposit fat rapidly and will usually have a body fat level of over 15 per cent by the time they are three weeks old. The fat serves as an energy reserve and helps them adapt to a reduction in milk intake and overcome the stress associated with weaning.

Males

The male reproductive system (Fig 7) is characterised by a pair of relatively large testes, which can weigh over 300 g each in some European or American breeds. Together with the secretions from the accessory sex glands, the testes can produce up to a litre of semen in a single ejaculate. To facilitate the transfer of such large quantities of semen at coitus, the end of the penis is spiral in shape, which enables it to lock into the cervix of the sow. The duration of coitus varies but may last for 20 minutes.

Puberty, or the ability of the boar to serve a sow, generally occurs at around four months of age, but may be earlier in unimproved breeds. However, boars should not normally be used until they are seven months old. Young boars are susceptible to bullying by mature sows, and this may adversely affect their subsequent mating performance.

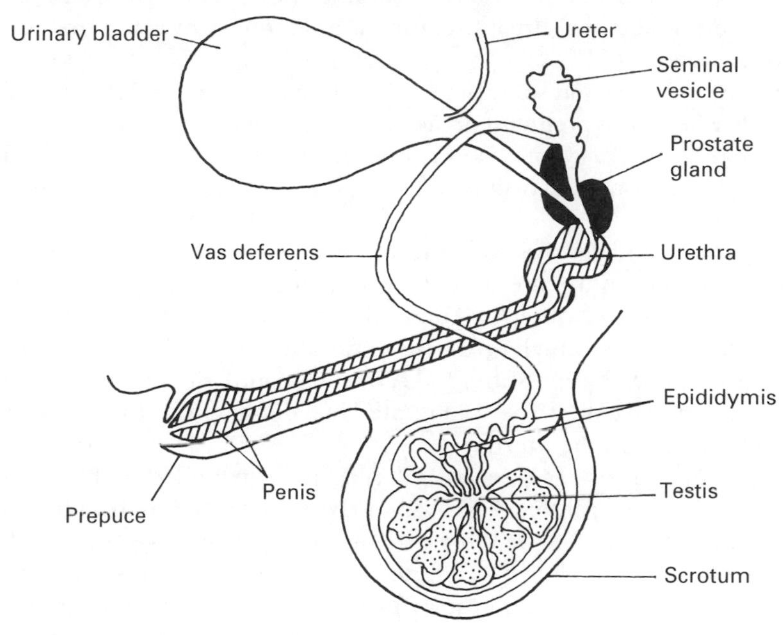

Fig 7 *The male reproductive system*

Females

The female reproductive tract is distinguished from other livestock species by the long, convoluted uterine horns (700–800 mm in length), which are designed to accommodate large numbers of foetuses (Fig 8). The sow will ovulate simultaneously from both ovaries, normally shedding

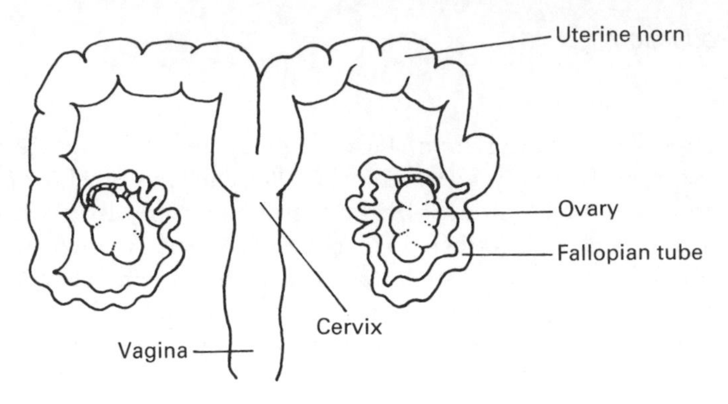

Fig 8 *The female reproductive system*

between 11 and 24 eggs. Puberty, marked by the onset of oestrous cycles, occurs at between five and seven months, but may be as early as three months in unimproved breeds. The number of eggs shed at ovulation, and therefore potential litter size, increases gradually over the first few oestrous cycles.

The sow will cycle and show heat every 21 days (range 18–24). She will not cycle when she is pregnant or lactating, although sows will sometimes show heat during lactation when run in groups. Heat lasts from one to three days, and ovulation occurs by the second day of oestrus or any time thereafter.

After coitus and fertilisation, the embryos space themselves evenly throughout the entire uterus before implantation. Competition for space, nutrients and other unknown factors often results in uneven growth rates *in utero,* which gives rise to variation in piglet birth weight. The lighter pigs will then suffer a disadvantage in their early post-natal life. This problem tends to be accentuated in older sows, due to the effects of wear and tear on the uterus.

Pregnancy lasts for 114 days but will tend to be extended slightly with sows carrying smaller litters. Farrowing may vary in duration from two to 24 hours and will tend to be longer the more piglets that are produced. However, due to the relative difference in size between piglets and dam, and the type of placentation in sows, farrowing is normally a straightforward process. The incidence of stillborn piglets, which may be due either to death *in utero* or during the birth process, is greater in larger than in smaller litters.

Overall, reproduction in the sow results from a complex hormonal interplay between the brain, the pituitary gland, the ovaries and the uterus. These complex relationships must be borne in mind by the pig-keeper when designing management strategies, otherwise optimum reproductive performance by the sow will not be achieved.

Behaviour

Pigs are not solitary animals and will generally benefit from social contact with each other, even if only by sight or smell. (For more information see *Livestock Behaviour, Management and Welfare* in this series.) At the same time, groups of pigs will always establish a social hierarchy. This starts at birth, as the piglets compete for a teat position. When strange pigs are penned together, they will invariably fight, which can cause considerable stress and physical damage to individuals. Once settled, however, pigs will huddle together in order to retain body heat in cold weather.

In common with females of other species, sows are notably more docile during pregnancy than when they have just produced young. Just prior to farrowing, the sow will prepare a nest from her bedding. She may be irritable during this period and, if confined without access to bedding material, she may experience stress during the farrowing process.

Contrary to popular belief, the pig is not a dirty animal and will normally defecate and urinate away from its resting or lying areas. However, this pattern tends to break down if pigs are overcrowded or stressed in other ways. When temperatures are high, pigs will often roll in their own faeces and urine in an attempt to increase evaporation and keep cool.

Recent studies have highlighted the importance of the interaction between pigs and humans in relation to productivity. If pigs live in fear of their stockman, their growth and reproductive performance are likely to be depressed.

Effect of climate and temperature

Due to its evolution within forest conditions, the pig has better mechanisms for retaining heat, especially a well-developed subcutaneous fat cover, than for losing heat from the body. The pig has sweat glands only on the snout and is therefore unable to dissipate much heat by sweating. Furthermore, the skin of certain breeds, e.g. Large White and Landrace, has no protection against the sun and these pigs need access to shade, or mud in which to wallow, or they may suffer from sunburn.

Like man, the pig is a homeotherm, and needs to maintain a constant deep body temperature. Nature has designed the metabolism of the pig to operate most effectively at 39ºC. For a certain range of environmental temperature, known as the zone of thermal neutrality (Fig 9), the pig can maintain the correct body temperature by varying blood flow to and from the skin. Interestingly, the extent of this zone changes quite markedly according to the weight of the pig (Fig 10).

At the lower end of this zone (point Y in Fig 9) is a critical temperature when the pig has to divert feed energy to increase heat production and

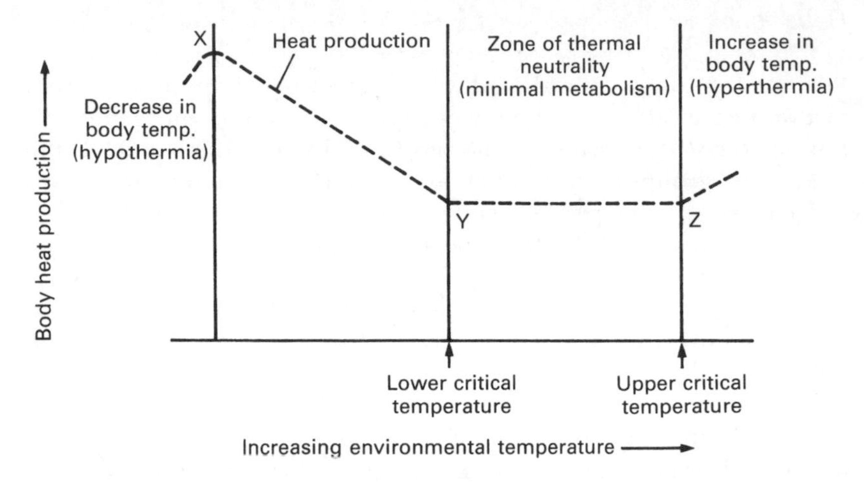

Fig 9 *Effect of environmental temperature on body temperature and heat production in the pig*

maintain its body temperature. The lower critical temperature will vary between pigs according to a number of factors. These include how fat (well-insulated) or thin the pig is, how much feed it is eating and therefore how fast it is growing, whether it has bedding to help prevent heat loss, whether it can huddle with pen-mates, and whether it can make postural

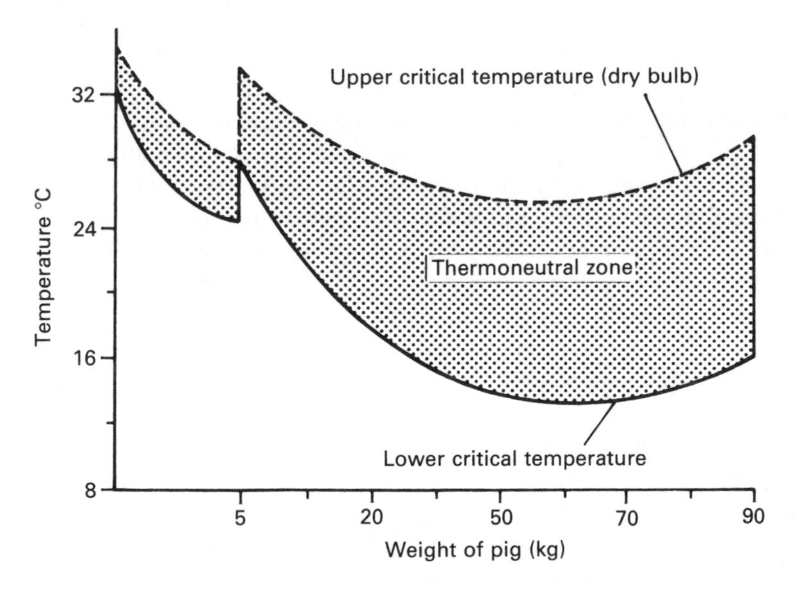

Fig 10 *Change in the extent of the thermoneutral zone with weight*

changes to minimise heat loss. Eventually, as the ambient temperature falls further, the pig can no longer maintain its body temperature in spite of high heat production, resulting in hypothermia and death (Point X).

Of greater interest in the tropics is the effect of rising temperature. When the environmental temperature approaches body temperature, the pig will attempt to increase evaporative heat loss by sweating (through its limited sweat glands), panting, postural and positional changes, and by wallowing in water, mud or excrement. In addition, it will reduce its energy output by decreasing its feed intake. However, as the means of dissipating heat in the pig are not very efficient, particularly if it cannot wallow, it will soon reach an upper critical temperature (Point Z). This is associated with hyperthermia and heat stress and the pig will die if the situation cannot be reversed.

The newborn piglet is particularly sensitive to low ambient temperatures. Pigs are born with virtually no subcutaneous fat cover and limited carbohydrate reserves and at birth they will suffer an immediate drop in body temperature. The weaker piglets may battle to obtain an adequate milk supply and, if they need feed energy to keep warm, they can very quickly develop hypoglycaemia (low blood sugar), and die of cold.

Stress

Stress factors can take many forms, involving fear, pain, temperature, direct sunlight, restraint, fatigue and interference with natural behaviour patterns. Stress will quickly lead to reduced performance and productivity, and specifically to gastric ulcers (just as in humans), greater susceptibility to infectious diseases and higher mortality rates. It is therefore of paramount importance that we understand what constitutes the major stress factors in pigs in different circumstances so that production systems can be designed to minimise these effects.

Immunity

The immune system acts to protect the pig from disease. As soon as disease organisms attempt to enter the body, the immune system is activated to fight and destroy the organisms. Either the immune system wins and the animal remains healthy, or the pig contracts the disease. Since the first edition of this book, a greater understanding has developed of the importance of the immune system in relation to productivity in pigs. As soon as the immune system is activated, the body diverts nutrients from productive processes (e.g. growth, lactation) to support the fight against disease. In southern Africa, it has been shown that pigs isolated from the main disease pathogens (known as specific pathogen-free pigs) show performance levels some 20 per cent higher than pigs reared in conventional systems. Management techniques suitable for the tropics that will reduce the disease challenge or enhance the immune system are considered in later chapters.

3 Systems of production

Small-scale systems

Pigs kept as scavengers

This is the traditional system of rearing pigs in large parts of the tropics, and is the simplest and cheapest to set up (Fig 11, Stage I). Each family, kraal or village keeps a few pigs, which are allowed to wander freely and feed when and where they can. They may receive supplementary feed if it is available and this is generally of low nutritional quality, such as banana and maize stalks, water hyacinth, rice bran, local herbaceous plants, by-products of beer-making or kitchen waste. Where pigs are particularly valued for ritual slaughter, as in parts of Asia, a few may be confined and fed for a 3–6-month fattening period prior to important ceremonies.

Indigenous breeds of pig predominate in this system, because they are adapted to the local environment and their relatively small size and high mobility render them best able to cope with the conditions (Fig 12). Productivity is normally low and feed supplies are often seasonal and/or erratic. The result is that sows breed irregularly, rates of piglet mortality are high and growth rates are low. Pigs raised on this system are particularly susceptible to infestation with parasites and invariably carry a heavy burden of intestinal roundworms and tapeworms. A particular hazard is that pigs can access sources of parasites, e.g. human excreta, which can then be transmitted back to man when he eats the meat (see Chapter 7).

The majority of scavenger pigs are owned by subsistence farmers and are not produced with any particular market in mind. Instead, pigs play an important socioeconomic role as a form of bank, being sold at times of cash shortages or unexpected needs in the family. The numbers kept are usually small, in the region of one to three breeding females per herd.

Semi-intensive production

In these systems, often known as 'backyard' systems, pigs are confined and there is a commitment on the part of the farmer to feed his pigs

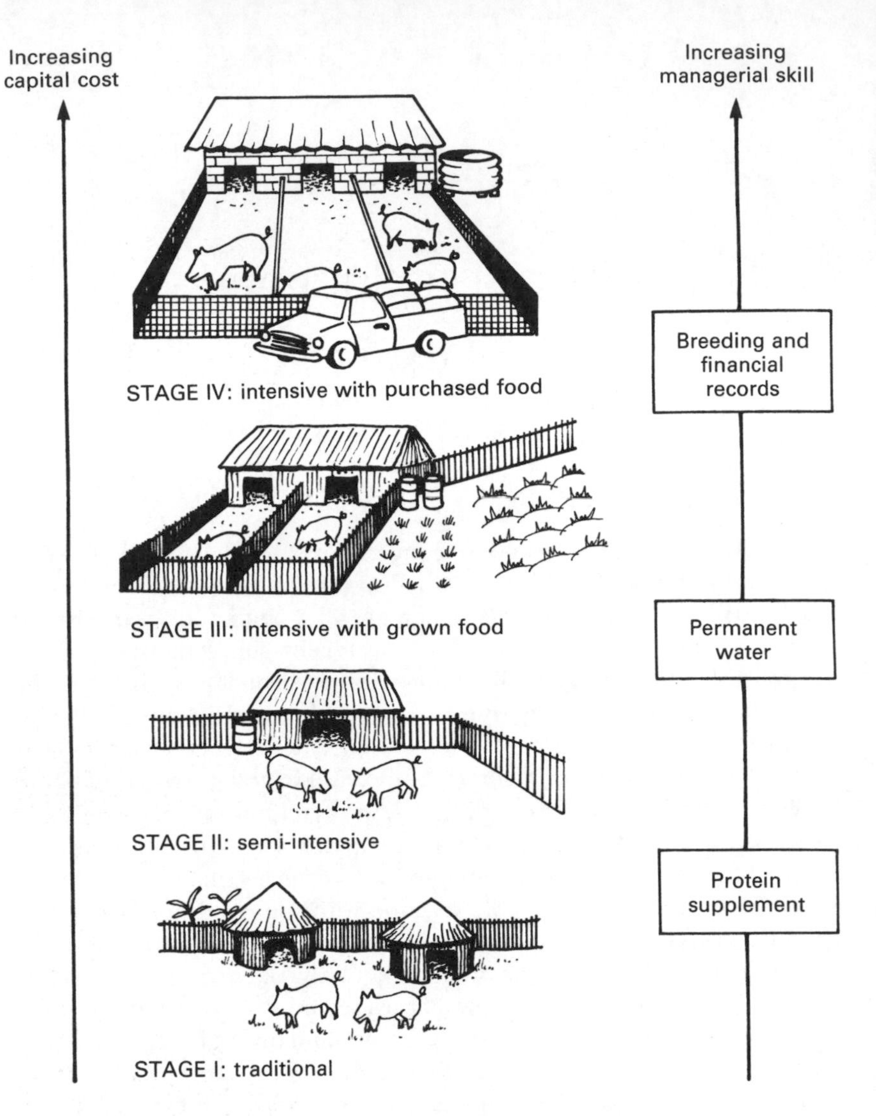

Fig 11 *The ladder of piggery development (Walters, 1981)*

(Fig 11, Stage II). Pens or sites are mainly of very simple construction (Fig 13). In South East Asia and West Africa, they may be made of bamboo and elevated; elsewhere the pigs may be tethered in larger yards or paddocks (Fig 14).

Feeding is based on kitchen waste, vegetables and by-products, and management is generally minimal. As a consequence, productivity tends to be relatively low and mortality can be high. Although local and indigenous pigs predominate, crosses between indigenous and exotic

Fig 12 *Indigenous pigs run on the scavenger system of production*

Fig 13 *Simply constructed pig pens used on semi-intensive systems*

breeds can be found in this system of production throughout the developing world. Marketing is largely indiscriminate and is directed by the immediate financial needs of the owner. Generally, herd sizes and productivity tend to be higher in these systems than in the scavenging systems.

Fig 14 *A tethered pig in Mexico*

Intensive production

Here, the small-scale producer has moved away from subsistence to more commercial production of pigs. Units may comprise up to 50 pigs, and the producer will grow and/or purchase feed specifically for his pig enterprise (Fig 11, Stages III and IV). The system of housing will be more sophisticated and will generally have a concrete or hard floor, adequate shelter, shade and pen space, and appropriate feed and watering facilities. In order to justify the increased capital cost, the farmer will attempt to manage his pigs to optimise output, including some veterinary protection against parasites and disease. The type of pig raised will tend to be mainly the higher performance exotic, or a cross between exotic and indigenous breeds. Marketing may be informal, through local butchers, or into the large-scale commercial sector, but in any event will be planned to bring in a regular income for the enterprise.

Large-scale systems

Intensive system

This is the most common system of large-scale production. Units are generally capital intensive and may involve sow herds from 40 up to 1000 head. Modern high-performance breeds of pig or hybrids are used, and provided as far as possible with optimum conditions of housing, feeding and management in order to ensure maximum output. Housing will often be designed specifically for different classes of stock and environmental conditions.

These units, especially the larger ones, are particularly amenable to integration with grain production and stock feed manufacturing operations on the one hand, and processing and marketing on the other. The pigs will invariably be marketed through a processor in order to maximise returns on the carcasses.

Extensive systems

There is a trend throughout Europe and America towards less intensive systems of production, particularly for sows. These systems, often known as 'outdoor' systems, entail keeping sows in paddocks and providing individual huts or arks for farrowing and shelter. Weaner pigs are generally raised under more intensive conditions. In Europe, hybrid sows, which have a greater ability to withstand climatic variation, are produced specifically for use on this type of system.

Production systems of this type can exist in the tropics and there is clearly potential for further expansion. The major advantages when compared with intensive systems are that less capital is required to establish the enterprise and sows can gain access to bulky feeds such as pastures, crop residues, cassava roots and sweet potatoes. In tropical regions, it is essential that adequate shade and wallows are provided. Moreover, there must be tight control of parasites and adequate fencing to prevent contact with endemic diseases, e.g. contact with the African bush pig and transfer of African swine fever.

Integrated systems

The integration of pig production with other ancillary enterprises is common in tropical Asia and involves various combinations including production of fish, algae, methane gas, ducks, water hyacinth and vegetables. Such bi- or tri-commodity operations enhance the efficiency of resource use and increase output for the overall operation.

By fertilising fish ponds with pig manure and effluent, algae are generated which can then be utilised by fish. As long as sufficient water is available for suitable fishponds, pigsties can either be constructed above the ponds, so that the manure can drop straight into the water, or close by, so that the effluent can be channelled into the ponds. Various *Tilapia* species are the most common fish, often mixed with small populations of carp (*Cyprinus spp*) and catfish (*Clarias spp*) or other predators. Fifty to 60 pigs produce sufficient effluent for one hectare of fishpond, which, if stocked at between 20 000 to 50 000 fish per hectare, can produce annual edible fish yields of 3.5 to 5 tonnes per hectare.

In some countries, people grow water hyacinth to harvest nutrients from the fishponds, and this can be fed back to the pigs. Alternatively, the nutrient-rich water can be used for irrigating vegetables, or ponds can be dried in rotation and vegetables grown in the dry pond beds. An alternative system, where other feed is available, is to use the pig effluent to produce algae, which can be harvested and dried and fed to pigs or other livestock. Sometimes, if the pig effluent is insufficient for a particular system, ducks are kept too, and their manure used as additional fertiliser for the ponds.

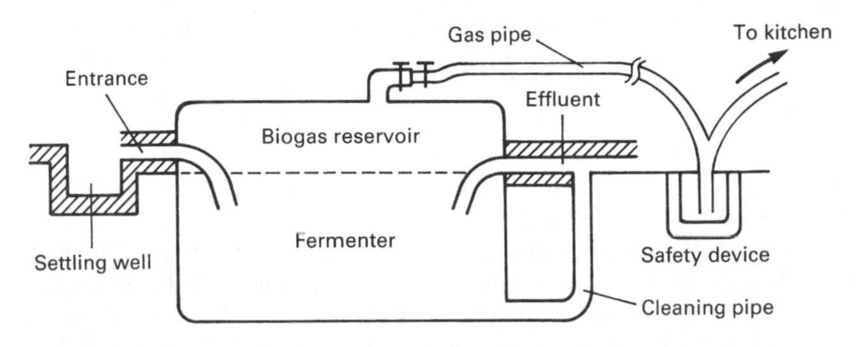

Fig 15 *A simple digester for methane production*

The solid fraction of pig manure can be used as an effective fertiliser for crops, particularly if it is properly composted, and this may prove to be the most cost-effective use of the by-product. The liquid effluent, however, is difficult to handle and use as a fertiliser for arable crops, and is generally wasted if not put to another use. An interesting development is the anaerobic fermentation of pig effluent for the production of methane gas. Relatively simple digesters can produce a steady source of methane, which can be used as a means of energy for domestic or agricultural use (Fig 15). These digesters are now in use in rural areas throughout the developing world, and it is estimated that seven pigs will provide enough dung to supply the domestic fuel needs of a family of five.

4 Breeds and breed improvement

Although some are small in number, there are over 90 recognised breeds and an estimated 230 varieties of pigs in the world. Pig breeds can be broadly classified into indigenous or unimproved types and the more modern exotic types, which have been selected and developed for specific commercial purposes.

Indigenous breeds (domesticated)

Most of the pigs kept in the tropical and developing world are of indigenous breeds, and a variety of shapes and sizes have evolved to suit a range of different environments. In general, these types of pigs are smaller and have shorter legs than exotic types (the mature weight of females being 40–120 kg), and they have the typical unimproved conformation of a large head, well-developed forequarters and relatively light hindquarters. This renders them more mobile and better able to forage and root for themselves. They are early sexually maturing and females may show first oestrus as early as three months of age. There are many variations of coat colour, but black and brown are the most common and white is infrequent. Some are hairless while others have relatively long hair. The main breeds and types found in the major regions of the tropics are listed below.

Africa

In many countries in Africa, pigs have not been characterised into specific breeds, and are variously referred to as 'indigenous', 'local' or 'unimproved' pigs. The situation is further confused in some areas by interbreeding with imported exotic strains.

In North Africa, pigs are not common and, where they do occur (e.g. Egypt, Morocco, Sudan and Tunisia), they are of extremely primitive type with long thin heads, small prick ears and slender poorly-muscled bodies. Similar primitive types have been reported in parts of West Africa, e.g. Congo (formerly Zaire) and Angola. In some parts of Africa, pigs are kept more specifically for meat production. The Bakosi in Cameroon and the Ashanti dwarf in Ghana are examples of such indigenous breeds. They are small (mature females reach only 40–60 kg), mainly black in colour and have prick ears.

'Unimproved' pigs are found in East, central and southern Africa. These are almost certainly descended from stock introduced by early European travellers, and therefore not truly indigenous, but they are relatively widespread. In Tanzania, some 60 per cent of the pig population

Fig 16 *Indigenous pigs in Zimbabwe, a) mature boar and b) sow and her litter*

is classified as indigenous, and some quite different types have evolved in various rural areas. The indigenous pigs of southern Africa have been classified by Mason and Maule (1960) into two distinct types: a) the Windsyner type, which is a long-nosed, razor-backed pig, common throughout Zimbabwe and found in parts of Mozambique and Zambia (Fig 16), and b) the Kolbroek, which is a short fat animal with a short snout and dished face and occurs in South Africa.

The productivity of African unimproved breeds is greatly influenced by their environment (Table 2). Interestingly, trials in Zimbabwe showed that, although litter sizes tended to be smaller, total live weight of the litter as a proportion of the weight of the sow at farrowing was of the same order as exotic sows (11 per cent). The efficiency of conversion of feed to bodyweight in indigenous sows is also equivalent to that of exotic types. Indigenous sows show excellent mothering ability, which results in very low piglet mortality without sophisticated housing. However, during the growth phase, growth rates and feed conversion efficiencies of indigenous pigs are less than those of their exotic counterparts (Table 3).

Table 2 Fertility and performance figures for indigenous sows in Africa

	Nigeria	Zimbabwe	S. Africa	Ghana (Ashanti Dwarf)
Litter size at birth	6.5	7.9	7.2	6.3
Litter size at weaning	5.5	7.5	–	–
Pre-weaning mortality (%)	15.0	5.0	–	–
Average weaning age (weeks)	9.0	8.0	8.0	8.0
Average weaning mass (kg)	–	7.6	9.0	7.0

Table 3 Comparison of growth rate and feed conversion efficiency of indigenous and Large White pigs fed a commercial diet from 8 to 32 weeks of age (Zimbabwe)

	Period (weeks)		
	8–16	17–24	25–32
Bodyweight gain (kg/day)			
Indigenous	0.28	0.51	0.45
Large White	0.44	0.58	0.68
FCE (kg feed per kg bodyweight gain)			
Indigenous	3.1	3.6	5.0
Large White	2.7	3.3	4.0

FCE = feed conversion efficiency
Source: Chigaru et al., 1981

China and South East Asia

The pig has been the main domesticated meat animal in China for centuries. It is estimated that there are over 100 different breeds, mainly derived from the *Sus vittatus* type. Epstein (1971) describes them in detail. In traditional Chinese agriculture, every small-scale farmer keeps one or two sows to dispose of farm by-products and household waste. The different breeds therefore have common features, including docility, an ability to digest and utilise low-quality diets and a tendency to carry a lot of fat. Chinese breeds are typified by a short dished head, a broad short body, often with a hollow back and pendulous belly, and short legs. Breeds kept in the south tend to be smaller than those found in central China. They are less prolific, but piglet mortality is very low. The Meishan breed has a relatively high prolificacy (in the region of eight extra pigs per sow per year) compared to European breeds and has been exported to developed countries, where it has been used in breeding programmes to increase litter size.

South East Asian pigs are derived from the Chinese breeds. In Thailand, for example, the local genotypes are believed to be derived from the Hainan breed of mainland China (Epstein, 1971). Research in the highland areas has shown typical litter sizes of 7.1 piglets born and 5.8 weaned per sow. The indigenous pigs of Malaysia and Singapore are very similar to the South China type, and other variants are found in the Philippines, Sarawak, Sri Lanka and Vietnam (Fig 17).

Fig 17 *Vietnamese pig*

South and Central America

Here, indigenous pigs are considerably more numerous than in Africa and, consequently, more breed types have been developed and described. Most are of the short 'lard' type, which probably originated from Chinese pigs introduced in the 15th or 16th Century. Fig 18 shows typical indigenous pigs in Haiti.

Fig 18 *Indigenous pigs in Haiti*

In Central America, the small, black hairless Pelon breed is common in Costa Rica, El Salvador and Mexico. Mature females weigh about 70 kg and their productivity is relatively low. Other variants of this type include the Yucatan pig, native to the Yucatan Peninsula, which is smaller, slate-grey in colour and tends to have a shorter snout. In Colombia and in parts of Bolivia and Venezuela, the black Criollo is a distinct breed, which has longish hair and a long, narrow snout. Pure Criollo pigs are now very rare in Bolivia due to indiscriminate breeding with imported exotic types. This interbreeding has resulted in a whole range of colour patterns in the village pigs. These crossbred Criollo pigs are also found in the Caribbean islands.

Pigs play a relatively important part in the economy of Brazil, and there are several indigenous breeds. These include the Pirapitinga, which is a small pig with very little hair, the Piau, which is black with white spots, the Pereira and the Canasta, which are lard types, black or grey in colour, with considerably more hair.

The Large White (Yorkshire)

The Large White was first developed in Yorkshire, England in the middle of the 19th Century, and has since become a very popular breed throughout the world (Fig 19). It is fast growing and strong framed with good length, and is renowned for its strength of leg. Females are prolific, good mothers and adapt well to confined conditions. The breed is widely distributed throughout the tropics, and is used extensively for crossbreeding. In Africa, for instance, the Large White x Landrace female is the most popular cross for commercial production. It is also used as grandparent stock in some of the main hybrids produced in Europe. In common with the Landrace, unless provided with adequate shade or wallows, the white skin renders it particularly susceptible to sunburn under tropical conditions. However, its white hair and skin render the carcass more acceptable to the consumer than that of the coloured breeds.

Fig 19 *Large White boar*

The Landrace

The Landrace originates from Scandinavia and is characterised by its forward-pointed lop ears (Fig 20). It was specifically developed for the bacon trade and typically possesses a long, smooth body with light shoulders and well-developed hams. It is a prolific breeder with excellent mothering qualities and produces lean, fast-growing progeny. The Landrace has a higher level of susceptibility to stress than some other breeds. Although not as numerous as the Large White, it is common throughout the tropics, and is highly favoured for crossbreeding purposes.

Fig 20 *Landrace boar*

The Duroc

This large breed was developed in the USA, although there have been suggestions that the British Tamworth breed was involved in the original stock (Fig 21). It is characterised by its deep red or rusty colour. The Duroc is fast growing and has been selected specifically for overall muscle and meat production. It can grow to a substantial weight without depositing too much fat. However, litter size and mothering ability are only average. It is claimed that the Duroc possesses a higher proportion of marbling fat (fat within the muscle) in the meat. An outstanding trait in the Duroc is its hardiness and resistance to stress and, consequently, lower levels of mortality. In tropical zones this is an important consideration, and the breed is increasing in popularity. In commercial production in parts of Africa, it is frequently used as a terminal sire (the final breed used in a crossbreeding programme) on white crossbred females.

Fig 21 *Duroc sows (Brazil)*

The Hampshire

The Hampshire is a medium size, black pig with a distinct white saddle encircling the forequarter. Originally a native of England, it has been developed as a modern breed in the USA. Hampshire sows are prolific, good mothers and produce above average quantities of milk. They are better able to cope with extensive conditions than white breeds. It is a meaty, well-muscled breed, which shows good efficiency of feed conversion. The Hampshire is very popular in crossbreeding programmes, both to produce a crossbred female and as a terminal sire.

The Berkshire

Although the Berkshire breed is on the decline worldwide, it remains popular in crossbreeding programmes in parts of the tropics. It is a relatively small, early-maturing pig, first developed in England for the pork trade. It has a black coat with characteristic white feet and nose. In the tropics, it has proved very hardy, and crosses well with indigenous stock. In some areas (e.g. Burma) where pig fat is used for cooking, it is valued for its high fat content.

Other breeds of interest

The Chester White
This breed was developed in America and has spread particularly into Central and South America. Females are highly prolific, but growth rates tend to be slow, and carcasses are shorter and fatter than average.

The Large Black
A hardy British breed, sows are very good mothers under extensive conditions. However, growth rates tend to be slow and carcasses are relatively fat. Although now rare as a purebred, there is evidence of the influence of the Large Black throughout the tropics.

The Middle White
An early-maturing breed similar to the Berkshire, it originated in England. It was used extensively for crossbreeding in South East Asia but is now rare.

The Pietrain
Of Belgian origin, this is a very lean and meaty pig, and is widely used in the production of modern hybrids. Introductions to the tropics have largely been unsuccessful due to the high level of susceptibility to stress. As an example, all the Pietrain pigs imported into Zimbabwe (then Rhodesia) in the 1960s died from heart failure as a result of one stress or another.

The Poland China

This breed of pig was one of the earliest to be developed in the USA. It tends to be large and fat and has been widely used in Central and South America. Improved strains of the breed are being developed.

The Tamworth

Characterised by its red colour, this is one of the oldest breeds of pig in England. The breed is exceptionally hardy, but is relatively slow maturing. In the past it has been very popular for crossbreeding purposes in tropical regions.

The British Saddleback

Another hardy British type produced by combining the Essex and Wessex Saddleback breeds. It has good milk production and mothering ability and is named after its distinctive markings of black coloration with a white saddle. In the UK, it has gained a new lease of life for the introduction of hardiness and mothering ability into hybrids used in outdoor production systems.

Genetic improvement

Genotype–environment interaction

When attempting genetic improvement of pigs in the tropics, it is important to realise that pigs selected for their superior performance in one environment will not necessarily be superior in a completely different environment. This is known as genotype–environment interaction. An example of this type of interaction is that restricted feed regimes will have a greater effect in reducing fat thickness in genetically fat pigs than in genetically thin pigs.

It follows that, if genotypes are selected under intensive conditions in a temperate environment and then transferred to the tropics, they need a modified environment (i.e. suitable housing, feeding and management) if they are to produce to their full capacity. In extreme cases, genotypes selected under intensive, temperate conditions have been expected to perform under tropical scavenging conditions. In such situations the animals have difficulty surviving, let alone growing and reproducing, and local or crossbred genotypes will be far superior under these conditions. Crossbreeds that have been successful in subtropical Africa include the Large White x indigenous (Fig 22) and the Duroc x indigenous (Fig 23).

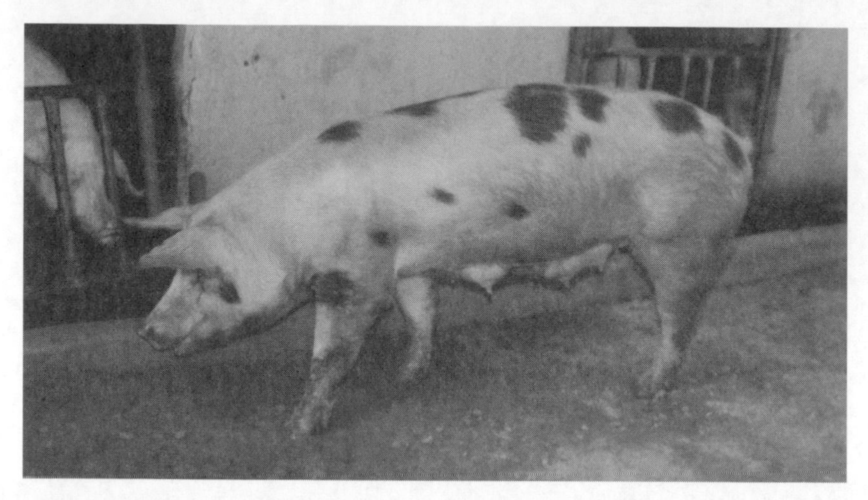

Fig 22 *Large White x indigenous sow*

Fig 23 *Duroc x indigenous gilt*

Methods of genetic improvement

The farmer has two main ways in which he can attempt to improve the performance of his pigs. He can either improve their environment or try to change their genetic make-up to increase their genetic potential. The various traits or characteristics of a pig are genetically controlled and inherited through genes containing the basic hereditary material. These genes can be manipulated to achieve genetic improvement by either increasing the frequency of favourable genes or combinations of genes by selection, or by introducing new genes into the population by cross-breeding with other breeds or strains.

During the last decade, tremendous progress has been made in the development of new technologies for genetic manipulation. These include gene transfer, which could considerably accelerate genetic improvement compared with conventional breeding. Such techniques could be particularly important in the tropical situation because they could help introduce useful traits (e.g. improved lean growth rate and carcass composition) into indigenous breeds without impairing the hardiness and adaptive characteristics they possess.

Selection for improvement

Genetic traits can be divided into simple traits governed by a single pair of genes, such as shape of ears or coat colour, or complex traits controlled by many genes, which include the performance traits such as growth rate, feed conversion efficiency and carcass quality. The genes for simple traits are normally dominant or recessive. If present in the heterozygote (i.e. a mixture of dominant and recessive genes), the dominant gene suppresses the expression of the recessive gene. Recessive traits will thus only appear when two recessive genes come together in the homozygous form (Fig 24). This means that the occurrence of a trait can be predicted. As an example, if a recessive trait is desirable (e.g. prick ears) then only prick-eared animals would be used as parents. This pattern of inheritance was first discovered by an Austrian monk called Mendel in his classic work with yellow and green peas, and is therefore known as Mendelian inheritance.

In the case of complex traits, the situation is entirely different. If we take growth rate as an example, within a given environment the individuals that possess genes favouring growth rate will exhibit superior growth rates compared to the rest of the population. Thus, if only animals with superior growth rates are selected as parents, this will increase the frequency of the favourable genes in the next generation. This is illustrated in Fig 25, where a population of pigs has been plotted according to growth rate, and only the animals growing faster than 750 g/day would be used as parents. The main factors affecting the efficiency and genetic progress of a selection programme are discussed below.

Definition of objectives

It is of paramount importance that selection objectives are clearly defined before a breeding programme is embarked upon, and that they are not subject to constant change. This is particularly important in tropical

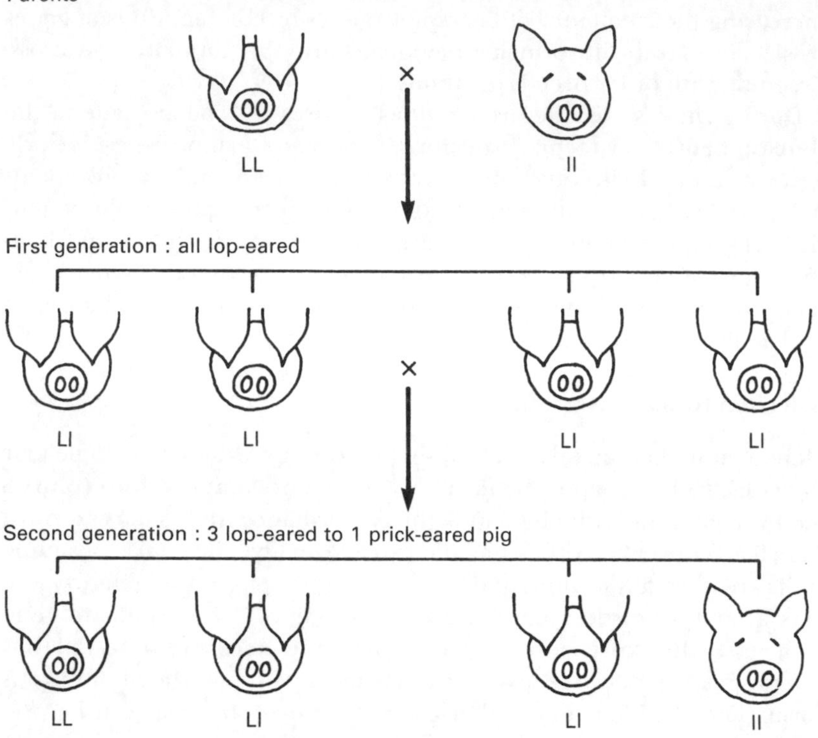

Fig 24 *Inheritance of simple traits: where lop ears (LL) are dominant over prick ears (ll)*

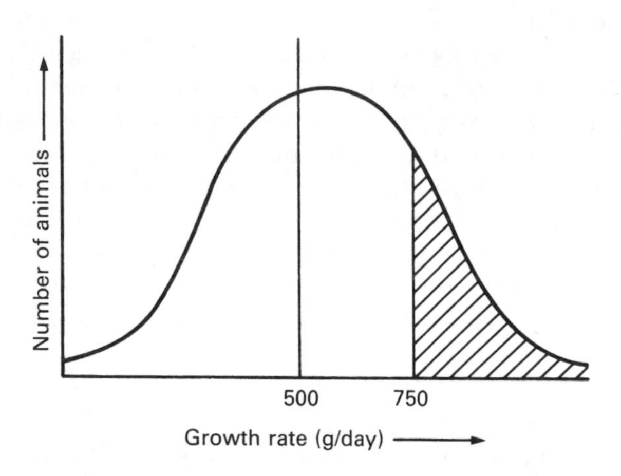

Fig 25 *Complex traits (e.g. growth rate): only superior animals within a population (shaded) are selected as parents*

conditions, where traits such as adaptability, coat colour and ability to produce on low-quality diets may be more critical than growth rates or carcass characteristics. Priorities may be completely reversed, e.g. fat pigs are preferred to thin pigs in some situations. The number of traits in a breeding programme must be kept to a minimum, because the more traits that are selected for simultaneously, the slower the progress for each trait. Nevertheless, selection priorities are bound to change with time under various pressures. The modern exotic breeds of pig have been selected for very lean carcasses and low levels of subcutaneous fat, but this has been achieved at the expense of meat quality. There is a definite relationship between leanness and marbling (intramuscular) fat, such that the leaner pigs have far less marbling. Marbling tends to make pig meat more juicy and flavoursome and consumer taste panels have consistently shown a preference for it. Disease resistance is also a high priority in genetic improvement programmes.

Selection differential

This is a measure of the superiority of the selected parents over the mean of the population from which they are derived. In Fig 25, for example, it represents the difference between the mean of animals in the shaded area (820 g/day) and the population mean (500 g/day). A greater difference illustrates further genetic progress. Clearly, a larger variation in a heritable trait in a population (i.e. a flatter curve in Fig 25) creates greater scope for selection differential.

Heritability

Heritability is a measure of a superior trait posessed by parents (compared with their contemporaries), and which is usually passed on to the offspring. More precisely, the heritability indicates the proportion of the total phenotypic variance that is due to additive genetic effects. Heritability estimates range from 0 to 100 per cent. Zero heritability implies that the trait is not heritable because there is no genetic variation. Conversely, a heritability of 100 per cent indicates that the trait is totally heritable whereby differences in environments between animals will not affect the phenotypic variation of such a trait. In general in the pig, reproductive traits tend to be of low heritability, growth traits of medium heritability, and carcass traits of high heritability (Table 4). Genetic progress will always be greater when selecting for traits of higher heritability.

Generation interval

This is defined as the average age of the parents when their offspring are born, and represents the time interval between generations. The shorter the generation interval, the more rapid the genetic progress and, if young boars are mated with gilts and replaced by selected progeny after one

Table 4 Heritability estimates for important traits in the pig

	Heritability (%)
Number of piglets born per litter	15
Litter size at weaning	7
Mean weaning weight	9
Daily weight gain	21–41
Feed conversion efficiency	20–49
Dressing-out percentage	26–40
Mean back-fat	43–74
Carcass length	40–87

litter, the generation turnover can be as short as one year. Even in the normal situation, where it is assumed that progeny born in first-to-fifth litters are equally likely to be chosen as replacements, pigs (with an average generation interval of 2–2.5 years) have a great advantage over other domestic meat-producing species such as sheep (3–4 years) and cattle (4–5 years).

Accuracy of measurement of traits
The success of a selection programme is entirely dependent on the accuracy of the records that are used. Some valuable traits are easy to measure, e.g. live weight gain, but others are more difficult to record accurately, e.g. feed conversion efficiencies and carcass measurements. Before embarking on a selection programme, it is essential to ascertain that the traits involved can be accurately measured.

Estimating the response to selection
Once the values for heritability (h^2) and selection differential (SD) and generation interval (GI) have been determined, the genetic gain per year can be estimated according to the formula:

$$\text{Genetic gain per year} = \frac{h^2 \times SD}{GI}$$

Testing procedures

Progeny testing

The boar has a relatively large influence on the characteristics that will be inherited by the next generation of the herd so testing systems have tended to concentrate on the male. Boar progeny testing systems have been in operation throughout the world for a long time, and are based on measuring the relative merit of a boar's progeny from several sows. This obviously gives a true indication of what a boar may be able to contribute towards the genetic improvement of a herd. However, in addition to being expensive, it takes a long time to accumulate sufficient data to evaluate a boar, and he will be relatively old by the time his potential is known. Progeny testing therefore has been largely superseded for traits of higher heritability by the performance test. Nevertheless progeny testing is useful for assessing traits that do not lend themselves to performance testing, e.g. sex associated and slaughter traits.

Performance testing

The basis of a performance test is that an animal's own performance is taken as a measure of its genetic merit. With traits of high heritability this can be used to predict how its progeny will perform. Thus the best individuals are selected from within a group of animals that have been treated similarly. The value and accuracy of a performance test can be checked by running a subsequent progeny test to see if the results agree with the merit order of the performance test. Performance tests can be carried out at central performance test stations, where the environment is standard for all animals tested. If facilities are adequate, testing can be carried out on-farm for within herd comparisons.

The traits and selection criteria used in a test will obviously vary according to their relative importance in different countries. They will also vary within a country depending on the use for which the pig is required. For example, different criteria will be used to select a boar for generating gilts for commercial breeding than a boar to be used as a terminal sire for the production of slaughter stock. Nevertheless, various combinations of growth rate, feed conversion efficiency and back fat thickness are the traits that generally form the basis for selection.

Selection methods

Independent culling levels and the selection index

There are several different selection methods; these are the two most commonly used.

In the independent culling method, a level of performance is set for each trait and if a pig fails to reach the desired standard in any trait it is automatically culled. It can be likened to an examination system where, if you fail any subject, you have failed the total examination. A major weakness of this technique is that if a pig has outstanding qualities in some traits, say growth rate and feed conversion efficiency, and just fails to reach the standard on conformation, it is culled. The genes for the outstanding traits are therefore lost. This method is the main system used for judging merit in pedigree breeding systems.

In the index method, the traits to be selected for are combined for one animal into a total score or index. Each trait is normally weighted according to its economic value and heritability, so that the highest index animal should yield the best financial return. These weightings can be adjusted as economic circumstances change. The advantage of the index method is that exceptional performance in one trait can balance out a weakness in another. In addition, if two traits are correlated so that improvement in one leads to a simultaneous decline in another, this can be allowed for in the weighting.

Improvement programmes

Here we describe the national pig improvement programme in Zimbabwe, which illustrates how testing and selection procedures can be integrated into a genetic improvement programme. The programme is based on a pyramidal structure (Fig 26). A small number of registered breeders, who have been assessed to have the best herds genetically in the country, send their pigs to a central performance testing station (Fig 27). Approved boars and gilts then pass out to selected multiplication herds, which have been assessed as the next best in terms of genetic merit, where their progeny is performance tested on the farm. Animals that pass this test are then available to commercial producers, thus ensuring a constant supply of quality replacement stock to commercial herds.

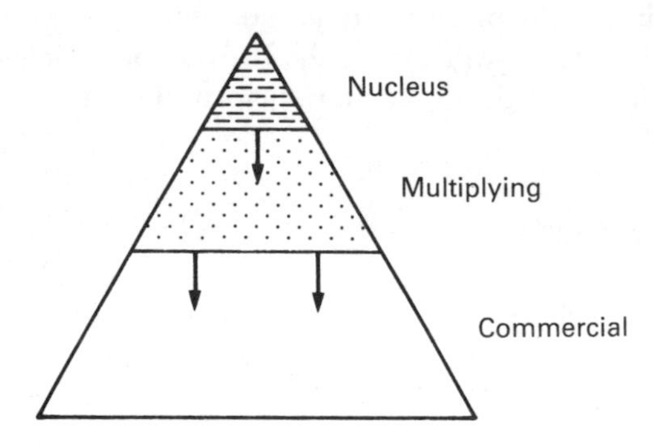

Fig 26 *The pyramidal structure of breed improvement in Zimbabwe*

At the central testing station, pigs are housed in pens of two and fed individually (see Fig 28). Their performance is measured as they grow

Fig 27 *The central performance testing station in Zimbabwe*

Fig 28 *Young gilts on performance test*

from 35 to 86 kg bodyweight. At the end of the test, the feed conversion efficiency is calculated and the fat thickness is measured by an ultrasonic machine. Pigs are assessed on the basis of an index derived from the relative economic values of feed conversion efficiency and fat thickness. Animals that do not reach the appropriate threshold index are culled. In addition, animals with any abnormal sexual characteristics, blind or insufficient teats, or other defects such as weak legs or undesirable body conformation, are also culled. On-farm performance tests are then carried out on the sons and daughters of approved centrally tested sires. Pigs are inspected for any weaknesses and, if they attain certain performance standards in terms of growth and fat thickness, they are made available for sale.

The potential value of regular performance testing can be seen from the results of a control Landrace herd, in which both boars and replacement gilts were selected on the above scheme over eight generations (Fig 29). The improvement in feed conversion efficiency represents a considerable saving in costs.

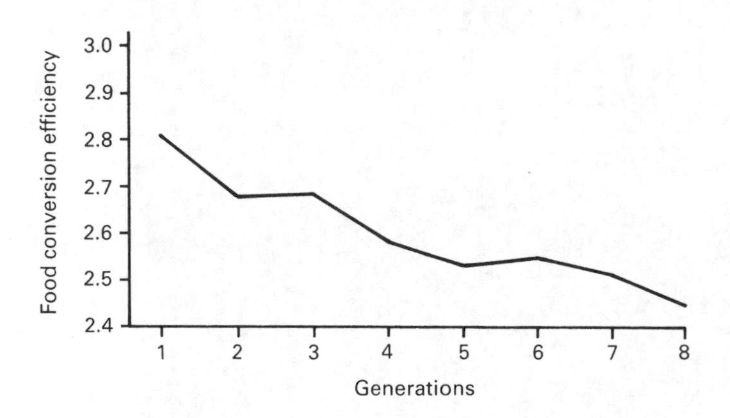

Fig 29 *Improvement in feed conversion efficiency as a result of central performance testing in a control Landrace herd for eight generations in Zimbabwe*

Practical crossbreeding

Crossbreeding, i.e. mating individuals from two separate breeds or strains, is another means by which the producer can attempt to genetically improve the performance of his pigs. The main advantages of crossbreeding are: a) exploitation of the phenomenon of 'heterosis', which occurs when two genetically different breeds are crossed; if the crossbred individual shows an improvement in performance above the mean of both parents it is said to exhibit hybrid vigour, and b) complementarity, or the ability to combine in one or more individuals the desirable characteristics of two or more existing breeds.

The degree of improvement that may be expected from heterosis is shown in Table 5. Higher levels of heterosis tend to occur in traits that are of lower heritability, e.g. reproductive traits. The degree of heterosis will always tend to be greatest when there is most genetic divergence between a tropical indigenous and an imported exotic breed. For this reason, the level of heterosis will always be highest in the first (F_1) cross, and decrease in subsequent crosses as genetic differences decline. However, the overall economic gain that can be obtained from heterosis effects can be cumulative within a crossbreeding system. Thus for litter size at weaning, for example, if the first cross dam gives a heterosis improvement of 11 per cent, and there is a further 6 per cent to be gained from her progeny's performance, the cumulative effect is 17 per cent, or more than one additional pig weaned per litter.

Table 5 Estimated levels of heterosis for some traits of economic importance in pigs

	+ % of mid-parent value
Litter size at birth	2–8
Litter size at weaning	5–11
Total litter weight at weaning	10–12
Post-weaning growth	6–8
Carcass traits	0–5

A number of different crossbreeding strategies can be used, all of which harness different degrees of heterosis. Of these, the two breed 'criss-cross' system is probably the most appropriate for the small-scale producer in the tropics (Fig 30). If a third pure breed is available, this can be extended to a triple crossing system using a sire selected from each of three breeds in rotation. Maximum heterosis is obtained when there is a large genetic diversity between the parent breeds, and there is therefore considerable potential for crosses between exotic and indigenous tropical breeds. The best use of such crosses is likely to be in semi-intensive systems of production (see Chapter 3), where the hardiness and foraging ability of the indigenous pig will be complemented by the fast growth and improved performance of the exotic. Indeed, large numbers of crossbred animals can be found on such systems. There are two main problems: a) it can be difficult to keep exotic boars in the tropics, as they often fail to survive due to disease and management hazards, and b) lack of breeding control often results in indiscriminate interbreeding. A possible solution is to use formal boar-holding centres, such as those in parts of northern Nigeria, where small-scale farmers can take their indigenous sows to be mated with purebred exotic boars. Alternatively, private farmers with intensive

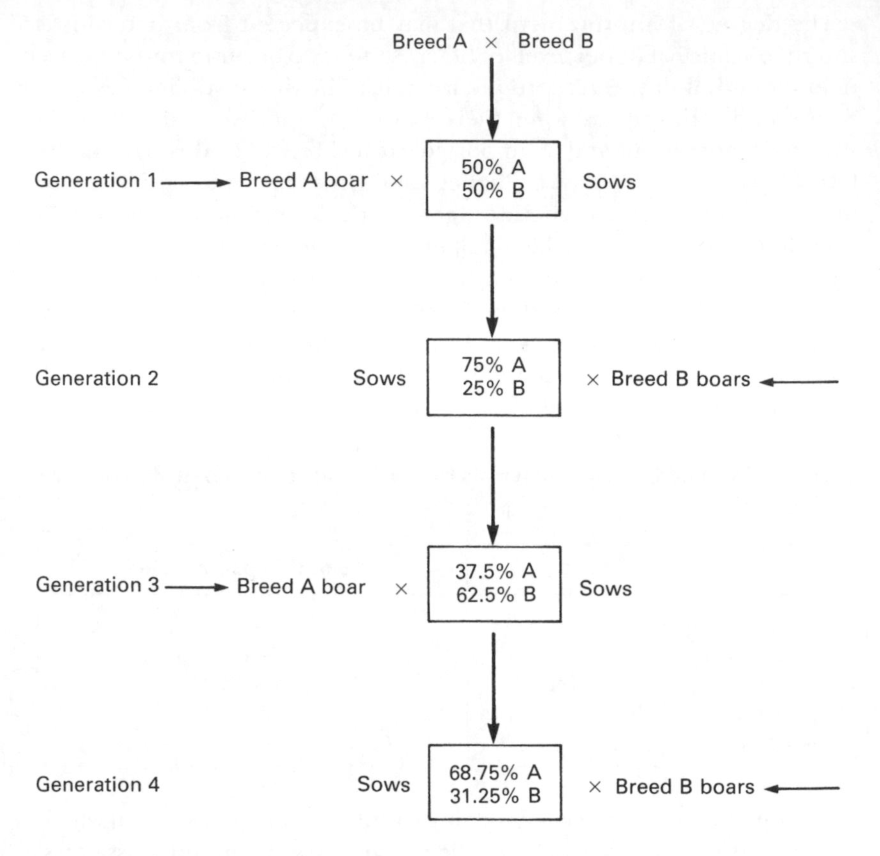

Fig 30 *Criss-crossing with two breeds of pig*

production units could be encouraged to produce crossbred progeny for sale to small-scale farmers. Similar methods could be used for generating exotic x indigenous boars for use on indigenous sows in small-scale production systems.

The benefits of heterosis in reproductive traits can also be exploited in boars. Crossbred boars have been shown to have greater libido, larger testes and higher sperm counts than their purebred contemporaries, leading to more reliable breeding and improved conception rates. This enhanced reproductive fitness is obviously likely to be a major advantage in extensive and semi-intensive pig production systems in developing countries.

The harnessing of heterosis is maximised in the genetic improvement schemes run by private breeding companies in the developed world. Each company maintains a network of super-nucleus and nucleus herds in which superior genes are concentrated. These herds are subjected to strong selection pressure with rigid and comprehensive testing of stock for important traits. Selected purebred stock from nucleus herds is then used to populate multiplier herds where improved lines and breeds are crossed. The resultant first-cross or hybrid gilts are sold to commercial producers.

Inherited defects

There are several anatomical defects that can affect the performance of pigs. Most are of genetic origin and therefore are likely to be passed on from one generation to the next. They may be simple (i.e. controlled by one pair of genes) or complex (controlled by several pairs of genes).

Simple defects

These can be eliminated from a herd by culling. The main defects in this category are:

- *Congenital tremors:* In this condition, piglets are born with rhythmic tremors of the head and limbs. It occurs at a frequency of about 0.1 per cent in the population. The recessive gene is carried by the dam and in carriers 50 per cent of male piglets are affected. Dams of affected litters should be culled.
- *Club foot:* This defect causes a deformed and swollen foot. It occurs only rarely and is restricted to the Landrace breed. Both male and female parents should be culled.

Complex defects

These are more difficult to eliminate, but the best approach is to cull affected boars. However, the incidence of these defects is generally relatively low, so it may be more economic in the long run to keep a good boar, even if it carries a known defect, rather than replace it with a genetically inferior boar that is not a carrier. If good records are available, the cost of the defect can be set against the gain in pig performance attributable to the carrier boar. Such a boar should only be used for the production of slaughter stock and none of his progeny should be retained for breeding. The main defects in this category are:

- *Scrotal hernia*: A weakness in the body wall allows part of the intestines to pass out into the scrotum. It occurs at a frequency of around 0.7 per cent and has an estimated heritability of 15 per cent.
- *Umbilical hernia:* A similar condition that occurs at the site of the umbilicus. Found at a frequency of around 0.6 per cent.
- *Imperforate anus*: The incidence of this condition is around 0.35 per cent. Mortality in male pigs is always 100 per cent, but often around 50 per cent of females survive as the faeces are voided via the vagina.
- *Splaylegs*: Piglets are born with either the front or hind legs splayed, sometimes both, and are unable to stand. The incidence is around

1.5 per cent, and the condition can be worsened by nutritional deficiencies. If piglets survive for three days, the condition often tends to disappear.

- *Hermaphroditism*: In this condition, females tend to exhibit male characteristics. Incidence is estimated at 0.07 per cent.
- *Cryptorchidism*: Also known as a 'rig' pig, one testicle in the male fails to descend into the scrotal sac. Found at a frequency of 0.22 per cent.
- *Female genital defects e.g. 'inverted nipples'*: These occur at a frequency of around 0.15 per cent.

Stress susceptibility – the 'halothane gene'

Throughout the world, many breeds of exotic pigs exhibit a condition known as porcine stress syndrome (PSS), which renders the animal particularly susceptible to stress. When an affected animal is subject to stress, for example through mixing with strange pigs, transport, exercise, fighting or mating, the pig may suffer an irreversible rise in temperature followed by sudden death. Slaughtered pigs will show a high incidence of pale soft exudative (PSE) meat, which is less attractive to the consumer than normal pork.

The condition is controlled by a single pair of genes that behave in a Mendelian fashion. That is, the pig exhibits PSS when two recessive genes come together according to Mendelian laws, as demonstrated for the prick ear trait in Fig 24. The ability to monitor the presence of this recessive gene in a population has been facilitated by the further discovery that when subjected to halothane gas anaesthesia, PSS pigs that are homozygous for the recessive gene become rigid and hyperthermic, whereas stress-resistant pigs (that are either heterozygous or do not possess the recessive gene at all) become relaxed. Thus it has become known as the 'halothane gene'. In different breeds, the incidence of a positive reaction to halothane anaesthesia has varied from zero in the Large White and Duroc to over 80 per cent in the Belgian Landrace and Dutch Pietrain.

Pigs that are recessive for the halothane gene can easily be identified (using halothane anaesthesia) and culled from the herd. However, the heterozygous condition confers some advantages, such as good muscling ability and leanness, on the pig. The advantages of the heterozygote can be exploited by using halothane reactor boars as terminal sires on female lines that are free from the gene. The progeny should, of course, be raised for slaughter only and not used in breeding replacement stock.

Artificial insemination

Artificial insemination (AI) involves the collection of semen from a boar and later introduction of semen into a sow or gilt by means of a catheter.

This contrasts with natural service where a boar mounts a sow and ejaculates his semen. The major advantage of AI is that it allows for wider use and distribution of semen from high genetic merit boars. The ejaculate from one boar can be used to inseminate up to 25 sows. Recent advances in boar semen storage make it possible for developing countries in the tropics to import the very top genetic stock from developed countries (e.g. in the UK, only the top 5 per cent of boars performance tested by the Meat and Livestock Commission are eligible for entry to AI studs). This calibre of genetic material would not otherwise be available to developing countries. Additional benefits of AI:

- Overcomes the need to purchase, house and feed a boar. This is particularly pertinent to the small-scale producer who cannot justify keeping a boar for a small number of sows, and who cannot afford a boar of good quality. The effective use of AI is especially relevant where small-scale producers are involved in group or co-operative pig development schemes, when their units can be serviced from a central boar holding centre.
- Prevents the transmission of disease from farm to farm by the sale and purchase of boars and by boar to sow contact.
- Overcomes the practical problems of differences in size of males and females. On occasions, this problem can severely limit the use of boars of high calibre.
- Reduces the risk to stockmen of handling boars for natural service.

Semen collection

Although various artificial vaginas and electro-ejaculators are available, they are not necessary for successful semen collection. Boars can be easily trained to mount a dummy sow device or an oestrous sow, and firm pressure on the penis with a gloved hand causes ejaculation to occur. The first low-sperm fraction of the ejaculate should be discarded, and the second sperm-rich fraction can be collected through a filter funnel, which removes the gelatinous fraction, into a warmed (30°C) bottle.

A drop of semen can be observed under a microscope to check its fertility characteristics and, if desired, the semen can then be diluted. A number of diluents and extenders are available, and the individual doses are normally stored in 50 ml plastic bottles for up to three days at 15–20°C. The number of spermatozoa used under commercial conditions for one insemination varies from 1×10^9 to 3×10^9.

Insemination

This involves the insertion of a rubber spiral catheter into the sow's vagina, and then rotating it in an anti-clockwise direction (Fig 31a) until the tip locks into the cervix. The bottle containing the dose can then be attached to the other end of the catheter and the semen introduced by gravity or slight pressure (Fig 31b). The insemination process may take up to 15 minutes.

Heat detection and timing of insemination

It is very important that conception rates achieved with AI approach those that occur with natural service. Accordingly, accurate heat detection must be carried out, preferably using a boar twice a day in order that the timing of insemination is correct (see 'Sows – mating management' in Chapter 8 for more information). To overcome inaccuracies in the detection

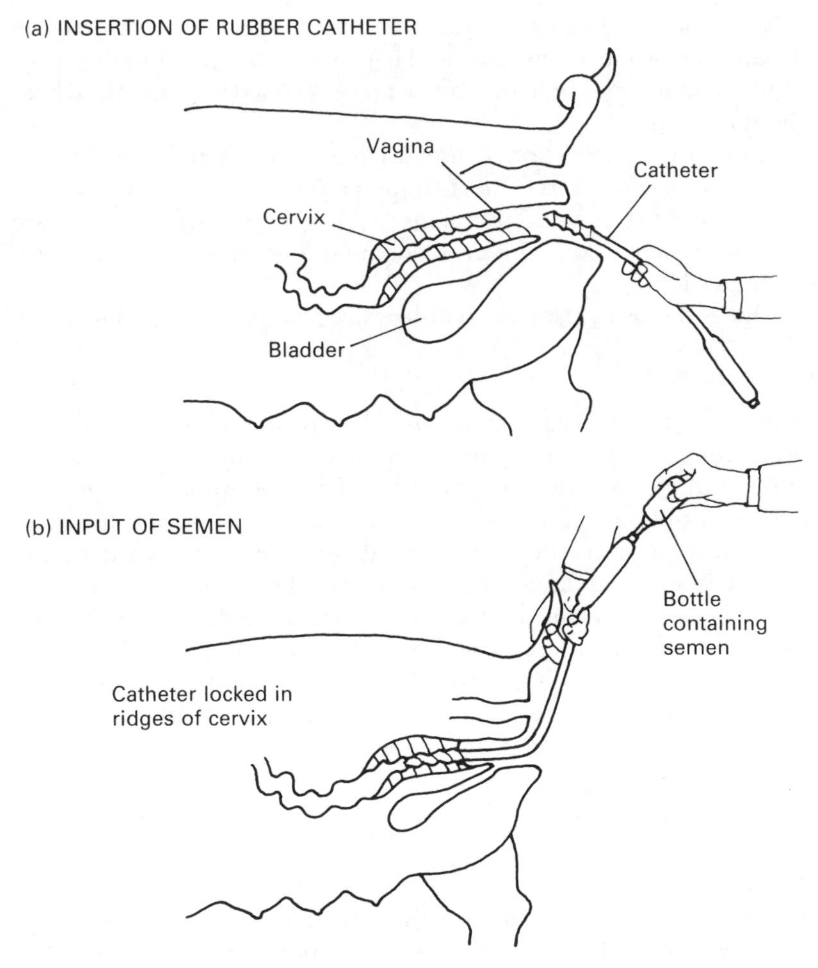

Fig 31 *Artificial insemination technique*

of the start of oestrus and the natural variations in time of ovulation, two inseminations approximately 12 hours apart are recommended (Fig 32). Some recently developed devices measure the electrical resistance of the vaginal mucosa. This changes in relation to hormonal events and can be used for accurate prediction of ovulation and hence the optimum timing of insemination.

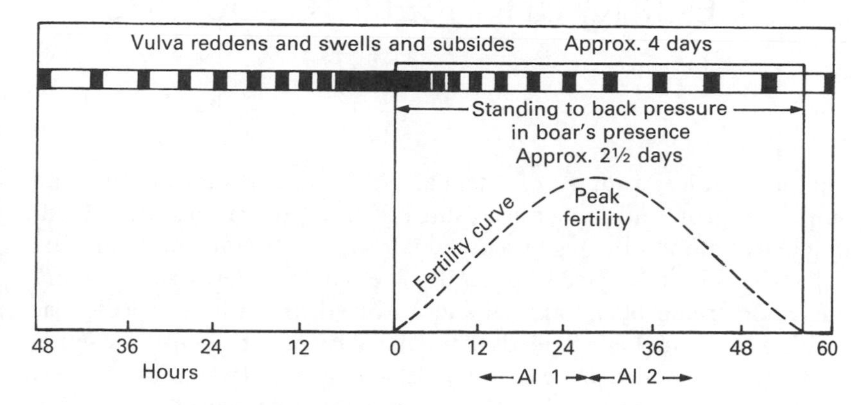

Fig 32 *Recommended times for insemination*

Semen storage

Unlike bull semen, boar semen can be damaged by the freezing and thawing process and, as a consequence, successful techniques for freezing boar semen have been developed only recently. Frozen semen can be obtained in pellet form or in straws, and will give acceptable conception rates. Being able to freeze boar semen has been a major breakthrough for the introduction of superior genetic material into developing country pig industries. Nevertheless, strict attention to detail in the handling of semen has to be adhered to if successful results are to be achieved. This involves thawing one insemination dose (10 ml of pellets) in a dry thawing box for exactly 3 minutes before incorporating into 45 ml of 50°C thawing diluent and inseminating. Boars vary in the 'freezability' of their sperm, and the semen of individual boars must therefore be screened before it is frozen.

Extenders have now been developed which allow fresh semen to be stored for up to seven days without a marked loss in fertility ('long life' semen). If transported by air, fresh semen can now reach developing country producers all around the world. This is the method of choice for routine pig AI as it will give higher conception rates and is less expensive than frozen semen.

5 Feeding and feed resources

The pig requires a supply of essential nutrients if it is to survive, maintain itself, grow and reproduce. Nutrients are present in various feeds, which are eaten by the pig and broken down into their component parts in the digestive tract (see Chapter 2). The basic nutrients then pass into the bloodstream of the animal and are used for various processes according to the biological needs of the pig. The major groups of essential nutrients are energy, protein, minerals, vitamins and water. Various feed additives can be given to improve the efficiency of conversion of nutrients into meat and milk.

Energy

Apart from water, sources of energy are the most important feed requirements of the pig and will most rapidly influence its survival if withdrawn. Energy can be defined as the capacity to do work and occurs in various inter-convertible forms such as chemical, thermal or radiant energy. It is normally measured in heat units (traditionally the calorie) and in modern livestock nutrition the megajoule (MJ) is the most commonly used unit (1 MJ = 0.239 Mcals).

Systems of measurement

The energy value of a feed is measured primarily as the Gross Energy (GE), which is the amount of heat produced when the feed is completely burnt in oxygen. Because not all this gross energy is available to the pig (i.e. it is not all digested), digestible energy is widely used to describe the energy content of pig feeds, where:

digestible energy (DE) = GE – energy loss in faeces

Energy can also be measured as Metabolisable Energy (ME), which allows for energy losses in the urine and combustible gases. However, as

ME is difficult to measure precisely and because it approximates fairly closely to the DE value (ME = 0.96 x DE), DE values are often used.

An alternative system of assessing the energy value of pig feeds is the Total Digestible Nutrient (TDN) system, which is widely used in the USA. The TDN is calculated as the sum of the digestible components of the diet, namely:

TDN = digestible crude protein
+ digestible crude fibre
+ nitrogen-free extract
+ (ether extract x 2.25)

One kg of TDN is equivalent to 18.49 MJ DE.

Requirements for maintenance

A pig is in a state of maintenance when its body composition remains constant and when it does not perform any work or give rise to any product (e.g. milk). The energy the pig requires for maintenance can be measured as the amount of heat given off by the animal, which represents its basal or fasting metabolism. Thus the total requirements of the pig for energy can be distinguished between those required for maintenance and those for production. In normal practical circumstances, an absolute maintenance situation (where the animal is in a state of equilibrium without gain or losses of fat or protein) is unlikely to occur. However, with the vagaries of feed supply in the tropics, adult pigs may spend periods of time close to a maintenance situation. Maintenance requirements are considerably affected by environment, particularly temperature, because both the retention and loss of heat require an energy input.

Although energy requirements for maintenance are related to body size, they do not increase as a fixed proportion of bodyweight, but become proportionately less in relation to bodyweight as the pig grows. In fact, maintenance requirements are also closely related to the surface area of the pig. This so called metabolic bodyweight is defined as bodyweight to the power 3/4 ($w^{0.75}$).

Requirements for production

Energy requirements for production are influenced mostly by the genotype of the pig; the environment has a relatively small effect. Therefore, rations must be designed according to the potential performance of the animal. In addition, energy requirements depend to a large extent on the type of production sought. For example, if a pig is expected to lay down fat, the energy costs will be high because fatty tissue contains only small amounts of water. On the other hand, if it is expected to deposit protein as lean meat, the energy costs are relatively low because lean deposits contain up to 75 per cent water. Other considerations apply when attempting to estimate the energy requirements for pregnancy and lactation. Pregnant sows use their food more efficiently than empty ones because of a phenomenon known as pregnancy anabolism. Lactating sows will often metabolise some of their own bodyweight and will lose weight if they are unable to eat enough food to meet the requirements for milk production.

Of the total energy available to the pig, obviously only the proportion that is surplus to requirements for maintenance is available for productive processes. The surplus is restricted by the pig's ability to eat, which declines in relative terms as the pig grows. Whereas a 15 kg pig can eat about four times the feed it needs to maintain itself, this reduces to three times or less at 90 kg live weight.

Fibre

As described in Chapter 2, the enzymes in the pig's digestive tract cannot digest fibre, which occurs to some extent in all plant material. However, the bacteria in the caecum can break down a small amount of fibre to fatty acids such as acetic, proprionic and lactic acid, which are then available as a source of energy. In general, a high level of fibre in the diet will reduce the availability of other energy sources, particularly if the fibre is not ground or milled. However, fibre is important in the diet as it promotes a healthy gut environment and ensures an efficient rate of passage of digesta.

Protein

Protein in the diet is broken down into amino acids, which are used to build up the essential organs of the body and skeletal muscle (lean tissue). Protein makes up some 15 to 20 per cent of the total bodyweight of the pig. Although the pig can manufacture some amino acids in the body, there are nine amino acids that cannot be synthesised and must be supplied in the diet. If they are present in insufficient quantities, the pig will not grow and may not even survive. These are known as the essential amino acids (Table 6).

Ideal protein and biological value

Feeding an appropriate balance of essential amino acids can be difficult, since pigs need different amounts of the various amino acids and different feeds have different amino acid contents. For example, valine and

Table 6 Amounts of essential amino acids in the 'ideal' protein for growing pigs

Essential amino acid	Ideal requirement for growing pigs (g/kg protein)
Lysine	70
Methionine and cystine	35
Threonine	42
Tryptophan	10
Isoleucine	38
Leucine	70
Histidine	23
Phenylalanine and tyrosine	67
Valine	49

Source: Agricultural Research Council, UK (1981)

isoleucine are in relatively plentiful supply in feedstuffs, whereas lysine and methionine are in shorter supply but are required in larger quantities by the pig. When the supply of essential amino acids in the diet is approximately the same ratio as that required by the pig, it is close to an 'ideal protein' from the pig's point of view. Table 6 lists the amino acid composition of an ideal protein source.

The biological value is a measure of how closely the protein ration approaches the ideal balance of essential amino acids. In practical terms, the total utilisable protein in a feed depends on the first limiting essential amino acid. For example, maize is often the main ingredient in pig diets in the tropics and, in this case, lysine is usually the first limiting amino acid. It is therefore important that when feeding maize to pigs, the ration is complemented by an additional source of protein that is relatively rich in lysine (e.g. soyabean meal). Groundnut meal is another common source of protein in the tropics – this has a higher total protein content but is less rich in lysine than soyabean meal (so of lower value for the pig).

Digestibility

The amount of amino acids available to the pig also depends on the digestibility of the protein. The main factors affecting protein digestibility are as follows:

- *Heat damage*: Digestibility of the protein will fall dramatically if the protein is damaged by heat, and this often occurs during the preparation of protein meals. Heat damage occurs when specific amino acids become bound to other compounds and are therefore unavailable. A prime example is lysine in blood meal where the theoretical lysine content is relatively high, but the 'available' lysine is low.

- *Degree of grinding:* If feeds are not ground properly, enzyme penetration and digestion is often impeded by indigestible plant material.
- *Poorly digestible carbohydrates:* If the proteins are contained within plant cells resistant to breakdown, the proteins cannot be released and will pass through the gut undigested.
- *Other anti-nutritional factors:* Other substances can prevent the digestion and uptake of amino acids by inhibiting the action of digestive enzymes, e.g. there is a trypsin inhibitor in unheated soyabean meal (see 'Anti-nutritional factors' on page 58).

Protein to energy ratios

Efficient utilisation of protein depends on the amount of energy available. Thus the amount of protein per unit of DE is more important than the absolute concentration of protein. The optimum ratio depends on:
- *The age of the pig:* The optimum protein to energy ratio changes steadily as the pig grows, being highest in the young animal and lower in the older pig, when protein requirements per kg live weight are less.
- *Genotype:* Exotic pigs, for example, need a higher protein to energy ratio than unimproved and indigenous pigs because they have a higher lean to fat ratio in their bodies (see Fig 6 in Chapter 2).

Maintenance, production and surplus

The first use of available protein in the body is to support maintenance requirements (as for energy). This involves tissue replacement and repair and manufacture of enzymes and hormones. In the growing animal, the maintenance requirement is usually less than one-third of the total protein need. The other two-thirds are needed for growth and the deposition of lean tissue, or for milk synthesis. Any surplus protein or amino acids that cannot be used by the pig are deaminated in the liver. The nitrogen fraction is excreted in the urine and the remaining carbohydrate is used as an energy source.

Minerals

Minerals are of vital importance in the diet of the pig. The growth and efficiency that should result from expensive inputs of protein and energy can be negated by small imbalances in or shortages of essential minerals. Defining the pig's requirement for minerals is complicated by the interaction that occurs between some minerals and the fact that not all the minerals in a feed are available to pigs because some cannot be digested. The pig's estimated mineral requirements are given in Appendix A and the essential minerals are listed on the following pages.

Calcium and phosphorus

Deficiency of calcium and/or phosphorus results in inadequate bone calcification, causing brittle bones that are easily fractured. This is expressed in younger pigs as rickets and as osteomalacia in mature pigs. If calcium or phosphorus are deficient in the diet of lactating sows, the sow will draw on supplies from her skeleton, resulting in lameness, rear leg paralysis and the 'downer sow syndrome', in which the sow becomes irreversibly crippled and unable to rise. It is important to maintain the right calcium to phosphorus ratio in the diet, which should not exceed 1.7:1 for pigs up to 20 kg live weight; 2.0:1 for 20–55 kg pigs and 2.4:1 for pigs over 55 kg.

Magnesium

Generally, there is adequate magnesium in the diet, especially in maize–soyabean meal rations. Deficiency leads to poor growth, bowed legs and a stiff-legged gait known as the 'stepping syndrome'.

Sodium and chlorine

These are treated together because it is common practice to add them to rations in the form of common salt (NaCl). Correct levels of salt are especially important in hot climates, because deficiency gives rise to poor appetite and low water consumption, leading to general ill thrift. Too much salt can also cause problems; if water is short, salt toxicity will occur at levels of 2 per cent salt in the dietary dry matter. In certain areas, particularly dry ones, water quality may be affected by high concentrations of salt, which increases the chance of toxicity and lowers water intake. In these areas, water may need to be treated before being offered to pigs. Toxicity is characterised by weakness, nervousness, convulsions, staggering and, ultimately, death. An appropriate level of salt in the diet is around 0.5 per cent.

Copper

Copper is required for the proper function of the nervous and immune systems. Copper deficiency results in nutritional anaemia and leg weakness. This is often associated with high levels of iron in the diet, as iron interacts with copper and reduces its availability. Pigs are tolerant of high copper levels – copper toxicity occurs at levels of 500 mg/kg DM in the diet, which is approximately 125 times the actual requirement. As a result, copper is often included in grower rations at well above the minimum requirements, where it acts as a growth promoter. This practice should not be encouraged because such high levels of copper are toxic to cattle and sheep. If copper-enriched pig feed were inadvertently fed to these species it could have harmful results. Care should also be taken with young pigs, as an inclusion rate of 300 mg/kg DM can cause toxicity, especially if zinc and iron levels are low.

Iron

Lack of iron causes anaemia, poor growth and a rough coat. It is a particular problem in piglets, because they are born with low iron reserves and there is very little iron in the sow's milk. As a consequence, piglets reared intensively without access to soil require a supplementary source of iron, either given orally or by injection (see Chapter 8). Piglets raised on extensive systems will normally obtain adequate iron by ingesting soil.

Zinc

Zinc plays a vital role in hormone secretion and helps maintain healthy tissue, thereby having a positive effect on the immune response to pathogens and the prevention of disease. It is also essential for sexual development and the formation of sperm in boars. Zinc deficiency gives rise to the condition known as parakeratosis. This causes red raised areas on the skin, which become crusty and may eventually coalesce over the whole body. Other symptoms include poor growth and reduced appetite.

Manganese

Manganese is also involved in stimulating immune function and reproduction. In growing pigs, manganese deficiency leads to lameness and deficient sows will have weak piglets. Deficiency is uncommon in normal pig diets.

Iodine

The main symptoms of iodine deficiency are goitre and an enlarged thyroid. Sows will produce weak, hairless pigs.

Selenium

Because of their interaction, selenium must be considered together with vitamin E: a lack of one will affect the utilisation of the other. There are three main conditions resulting from deficiency, namely: mulberry heart disease, hepatosis dietetica and muscular dystrophy. Mulberry heart disease normally affects pigs aged three weeks to four months of age, and the main symptom is sudden death due to cardiac failure. Post-mortem examinations will show lesions of the cardiac muscles. Hepatosis dietetica also occurs at between three weeks and four months of age and is characterised by a generalised oedema and death, with the ears turning blue. With muscular dystrophy, pigs develop a staggering gait and are unable to rise. It usually affects fast-growing pigs of between 30 and 60 kg live weight. Vitamin E is broken down by oxidation if stored at high temperatures. It therefore tends to be unstable under tropical conditions and, unless an antioxidant (such as ethyoxyquin) is added to the rations, a deficiency is likely to occur. This is particularly the case when selenium levels in the diet are low, as when maize–soyabean meal combinations are used as feeds, or when the levels of unsaturated fats in the diet are relatively high.

Chromium

Recently, chromium has been added to the list of important trace minerals in pig diets due to its role in the action of insulin and the reduction of stress. Several trials have shown that inclusion of chromium in the diet improves litter size and farrowing rate in sows and growth rate and carcass composition in grower/finisher pigs.

Mineral availability and interactions

Availability can be affected by the source of the mineral. For example, when phosphorus is supplied as phytate phosphorus, it remains bound to the calcium component and its availability is markedly reduced. Availability can also be affected by the metabolic state of the animal. Studies have shown that calcium and phosphorus availability will decrease with both age and size. For example, calcium availability declined from 85 per cent at 5 kg bodyweight to 45 per cent at 90 kg and phosphorus availability declined from 80 per cent at 5 kg to 65 per cent at 90 kg.

Traditionally, trace minerals have been provided to pigs in the form of inorganic salts in feedstuffs or as supplements. These are partially broken down during digestion and the availability of the element may vary considerably. A recent development has been to provide them in the organic or chelated form (also known as proteinates). In this case, the trace mineral is bonded to peptides or amino acids and can be absorbed via peptide or amino acid pathways, rather then through the normal mineral uptake pathways in the small intestine. The bioavailability and intestinal absorption is therefore enhanced. Equally importantly, the minerals are protected biochemically from adverse interactions with other minerals, which could reduce absorption and availability in the tissues.

Interactions can occur between minerals, between minerals and vitamins, and between minerals and other feed ingredients, where one ingredient can depress or enhance the utilisation of another. Some important interactions:

- calcium, phosphorus and vitamin D;
- vitamin E, selenium and fat levels;

- calcium and phosphorus influencing the metabolism of zinc;
- copper and zinc;
- copper and iron;
- iron, vitamin E and selenium.

Vitamins

Vitamins are organic compounds that function in small amounts (mg or μg) and are essential to the normal functioning of the body. They cannot be synthesised in adequate amounts by body tissues and, when lacking, they can provoke deficiency diseases. Estimated vitamin requirements for pigs are shown in Appendix A. Fourteen vitamins are normally considered to be required in the diet of pigs, of which four are fat soluble and ten are water soluble.

Fat soluble

- *Vitamin A:* Deficiency causes night blindness, birth of blind piglets, lacrimation of eyes, nasal discharges and a rough and dry coat. It also leads to a decrease in ovarian and testicular size and increased embryo mortality.
- *Vitamin D:* Through its relationship with calcium and phosphorus metabolism, deficiency can lead to rickets and osteomalacia. If pigs have access to sunlight, vitamin D may be synthesised in adequate amounts in the skin.
- *Vitamin E:* Interacts closely with selenium, and the main deficiency symptoms are listed under selenium. Deficiency will also lead to reproductive disturbances and reduced milk production.
- *Vitamin K:* Deficiency of vitamin K does not often occur, but can cause an increase in the clotting time of the blood, nasal bleeding and subcutaneous haemorrhages.

Water soluble

- *Thiamin:* Deficiency causes nervous disorders, cardiac lesions and high piglet mortality at birth.
- *Riboflavin:* Typical symptoms of riboflavin deficiency are a rough hair coat, loss of hair and dermatitis. Deficiency in sows leads to reabsorption of the foetus, premature farrowing and anoestrus in gilts.
- *Niacin:* Lack of niacin causes dermatitis of the skin, diarrhoea and anaemia.
- *Vitamin B6:* Slight deficiency will cause vomiting and diarrhoea, but more severe deprivation leads to disordered movements, lacrimation and blindness.

- *Pantothenic acid:* Deficiency causes scurfy skin and leads to the locomotion disturbance known as 'goose stepping'.
- *Biotin:* Lack of biotin leads to a rough hair coat, cracks in the feet, lameness and depressed reproductive efficiency.
- *Folic acid:* Lack of folic acid causes lesions in the mouth and buccal cavity, which renders the pig more susceptible to infection.
- *Vitamin B12:* This is an important vitamin in relation to general health, and a deficiency will cause diarrhoea, vomiting, rough coat, lack of co-ordination in the hind legs and reduced reproductive performance.
- *Choline:* A deficiency of choline causes splay legs in piglets and reduces piglet survival. It can also reduce reproductive performance.
- *Vitamin C:* Although not common, a lack of vitamin C in the diet will lead to weakness and haemorrhaging throughout the body. Vitamin C is often supplemented to improve the immune response and resistance to infection.

Vitamin availability and interactions

The availability of vitamins is influenced by a number of factors. For example, the B vitamin niacin is present in relatively large quantities in cereal grains, but is bound in a form that makes it unavailable to the pig. In addition, antagonists in pig feed can interfere with utilisation of this vitamin. In general, the vitamin potency of feeds will tend to decline during storage. The rate of decline depends on the conditions of storage, e.g. the intensity of light, ambient temperature, acidity, the presence or absence of interfering substances and the physical form of the diet. Because storage conditions in tropical areas are often far from ideal, it is very important to minimise the periods of storage of mixed feeds. Moreover, a synthetic antioxidant (e.g. methylene blue or ethyoxyquin) should be included when vitamin supplements are compounded as it will increase vitamin stability. As with minerals, the function and effectiveness of a vitamin may be influenced by interaction with other vitamins, minerals or feedstuffs.

Feed additives

These are designed to enhance gut health and improve the efficiency of feed utilisation. Their inclusion in diets depends on the cost-effectiveness of the response in pig performance. Traditionally, the feed additive market has been dominated by antibiotic growth promoters. However, the recent banning of antibiotics for this purpose in developed countries had led to radical changes in the range of feed additives. For example:

- *Organic acids:* Mainly lactic and fumaric acids, these serve to acidify the gut and stimulate the development of beneficial microbes.
- *Probiotics (mainly lactobacilli):* These also acidify the gut and stimulate the development of favourable bacterial populations in the gut.
- *Enzymes:* These supplement the digestive enzymes in the pig. They can digest carbohydrates (carbohydrases), proteins (proteases) and release phosphorus (phytases). Because pig feeds in the tropics tend to contain higher fibre and be of lower digestibility than pig feeds in temperate climates, there is likely to be potential for increased use of supplementary enzymes.
- *Oligosaccharides*: These bind to pathogenic bacteria in the gut and prevent them from attaching to the gut wall where they will cause irritation and disease.
- *Mycotoxin binders*: As tropical climates favour the development of fungi in crops at harvest, mycotoxins (particularly aflatoxins) can be a major poison hazard in pig feeds. Binders will detoxify the feed.
- *Herbs and spices (botanicals):* Many herbs have been found to regulate feed intake, stimulate digestive secretions and optimise digestive capacity. They provide essential oils and antioxidants, and stimulate the immune system. Rural communities in tropical areas often have extensive knowledge of the benefits of certain herbs for human well being and health and this could be applied to benefit pig production.

Water

With the exception of oxygen, water is the most important ingredient for the maintenance of life. Some 65 per cent of the body of a pig is water and water is involved in most of the chemical reactions that take place in the body. Water is of particular importance in the tropics and hotter and drier parts of the world, where it is often in short supply, and where the pig requires more water to enable it to maintain body temperature. Lack of water very quickly leads to a rise in body temperature and death. However, equally important is that sub-optimal amounts of water will have a major effect on feed intake and performance, and this applies irrespective of the system of production. It is thus essential in tropical climates that pigs have access to a supply of clean fresh water at all times.

Anti-nutritional factors

In the tropics, plant proteins are by far the most common source of protein for pigs. However, a number of these are associated with anti-nutritional factors in the form of toxins or other substances. These interfere with digestion and utilisation of ingredients, and it is essential that they are taken into account when formulating rations. A prime example is soyabean, which contains a trypsin inhibitor. This inhibits the action of trypsin in the pig's gut and can reduce the digestibility of protein to the extent that only 30 per cent of the protein in the ration will be digested.

Voluntary feed intake

The more the pig eats, the more nutrients are likely to be available above maintenance requirements for productive processes. Recent research has shown that when young pigs' appetites are increased above normal, the level of protein deposition and growth are markedly enhanced.

The pig will normally control its energy balance by regulating its voluntary feed intake. This is of particular relevance under tropical conditions where pigs often receive diets of low nutrient content and low energy density. Thus a pig will need to eat greater quantities of a diet of low nutrient density if it is to maintain its energy intake, but this will only occur up to a certain limit (Fig 33 point A). Thereafter, feed intake will not increase any further with dilution of the energy density, and energy intake will drop. A more detailed review of energy balance and voluntary feed intake can be found in Whittemore, Green and Knap (2001).

The art of feeding pigs in the tropics is to maximise use of cheaper feed sources of lower energy density while maintaining adequate energy intakes. The position is also complicated by the fact that high ambient temperatures and irregular water supplies will serve to further reduce the

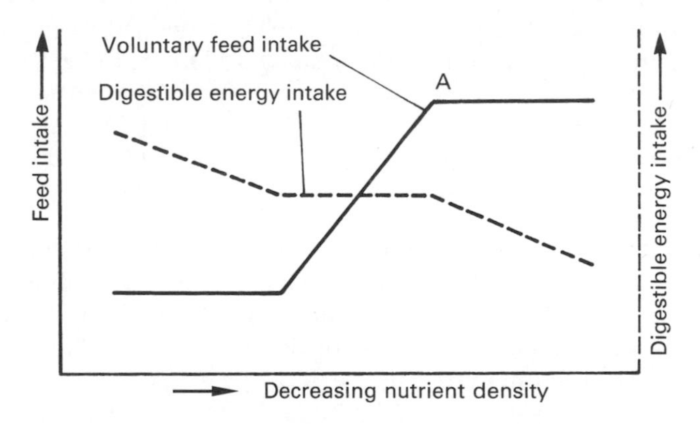

Fig 33 *Relationship between the concentration of DE in the diet and the total daily feed intake (after Cole, Hardy and Lewis, 1972)*

pig's voluntary feed intake. Palatability of the feed source must also be considered, because a high protein content in a feed is of no use to the pig if it cannot be induced to eat it.

Formulating diets

The first step in formulating a ration is to define the nutritional needs of the pigs. Estimates of nutrient requirements for pigs are available from various sources (see Appendix A). Secondly, available raw materials need to be listed together with their nutrient composition. The skill is then to consider the factors discussed above and to combine the ingredients into a ration that will meet the pigs' requirements at the lowest cost. A margin of safety should be allowed to cover inexact information about ingredients. Moreover, it is very important in a tropical environment to maintain a flexible approach to formulating a ration according to resources and conditions. Published estimates of requirements should be considered as guidelines only, and should be modified according to pig performance. In economic terms, providing for maximum performance of the pig may not be the most profitable course of action.

Formulating diets by hand or using a calculator

Diets can be formulated by hand calculations, where proportions of ingredients are varied on a trial and error basis until the desired nutrient requirements are obtained. The procedure for formulating a basic ration can be illustrated by a simple example. First, list the feed ingredients that are available, noting any constraints (Table 7). These details can be obtained from Appendix B. Then, if a grower ration is required, list the nutrient requirements for growing pigs from Appendix A (Table 8).

The approximate proportions of ingredients needed to meet these requirements, based on conventional diets, are known to be:

Energy sources	65.0–75.0 %
Protein sources	20.0–25.0 %
Calcium/phosphorus sources	2.0–3.0 %
Mineral/vitamin additives + salt	1.5–2.0 %

A first attempt to formulate a suitable diet for grower pigs is shown in Table 9. Note the amount of nutrient each ingredient contributes to the final ration is calculated on a percentage basis, e.g. maize is 72 per cent of the ration and contains 13.9 MJ/kg of digestible energy. Therefore, the amount of energy from maize is:

$$13.9 \times \frac{72}{100} = 10.0 \text{ MJ/kg}$$

Table 7 Nutrient composition of some typical feed ingredients

	Digestible energy MJ/kg	Crude protein %	Calcium %	Phosphorus %
Maize	13.9	8.9	0.01	0.25
Soyabean meal	12.5	46.0	0.25	0.60
Chickpea meal[1]	10.8	20.1	0.26	0.32
Dicalcium phosphate	–	–	22.0	9.0

[1] Chickpea meal is only available in small quantities

Table 8 Estimated nutrient requirements for a growing pig (bodyweight 10–20 kg)

Digestible energy MJ/kg	13.0
Crude protein %	19.0
Calcium %	0.7
Phosphorus %	0.6

Table 9 A possible ration for grower pigs

	Content (kg/tonne)	Digestible energy MJ/kg	Crude protein %	Calcium %	Phosphorus %
Maize	720	10.0	6.4	0.007	0.18
Soyabean meal	210	2.6	9.6	0.05	0.13
Chickpea meal	30	0.3	0.6	0.007	0.009
Dicalcium phosphate	25	–	–	0.55	0.23
Minerals/ vitamins + salt	15	–	–	–	–
Total	1000	12.9	16.6	0.61	0.55

This first attempt at formulating a ration shows that in comparison with the estimated requirements for growing pigs (Table 8), the energy content is slightly low, the protein content is too high and the calcium content is also slightly low.

Because protein is likely to be the most expensive ingredient, it is important to get the protein level correct. A useful technique for balancing the protein content is known as Pearson's Square (Fig 34). The protein level required (i.e. 16 per cent) is placed in the middle of the square, and the percentage protein content of the ingredient feeds go in the two left-hand corners. By subtracting diagonally across the square (16 – 8.9 = 7.1 and 42.8 – 16 = 26.8 in Fig 34) the proportion of ingredients required to give a 16 per cent protein content in the final diet can be read from the right-hand side, i.e. 7.1 per cent soyabeans and chickpea to 26.8 per cent maize (equivalent to 1:3.77). Rounding off the figures gives the proportions for a practical ration (Table 10), which contains the correct balance for a growing pig. Suitable mineral/ vitamin pre-mixes should be purchased and added to the ration according to estimated requirements (see Appendix A).

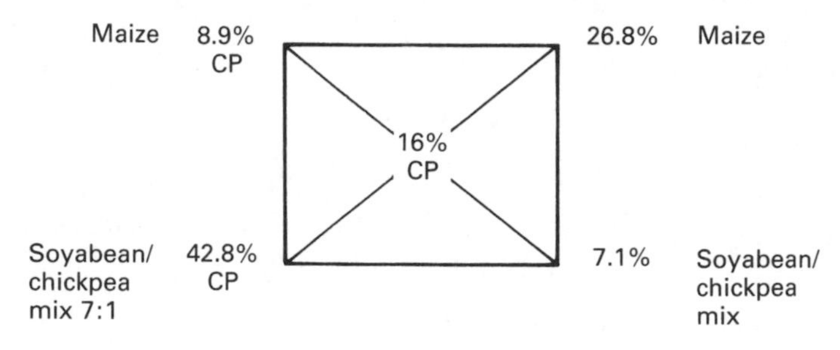

Fig 34 *Pearson's Square technique for calculating protein levels in diets*

Table 10 A corrected ration for growing pigs

	Content (kg/tonne)	Digestible energy MJ/kg	Crude protein %	Calcium %	Phosphorus %
Maize	750	10.4	6.7	0.007	0.19
Soyabean meal	184	2.3	8.6	0.06	0.11
Chickpea meal	26	0.3	0.5	0.007	0.008
Dicalcium phosphate	26	–	–	0.572	0.24
Mineral/ vitamins + salt	14	–	–	–	–
Total	1000	13.0	15.8	0.65	0.55

Computer formulation

Ration formulation is ideally suited to computer programming, and most feed companies and larger operations now formulate this way. Computer software can capture a database of available ingredients and their costs and can rapidly solve a whole series of trial and error calculations to arrive at the required diet specifications at minimum cost. It is imperative, however, that least-cost ration programming is carried out by a nutritionist, who can check that the end result makes nutritional common sense. The cheapest ration is of no use if the pig finds it unpalatable! Computer technology has developed rapidly during the past decade, with smaller, faster machines and software programmes that are very 'user friendly'. Most pig nutrition consultants will carry a laptop with them and will produce formulations on farm visits. This is particularly useful for small-scale producers and development schemes in remote areas. (*Poultry* in this series has more information about computer formulation.)

Feed resources

Feedstuffs can be classified into energy or protein sources according to whether they are rich in energy or protein. However, energy sources will contain some protein. Energy sources normally form the major part of the diet and are therefore required in larger quantities and will exert a greater influence on overall cost of the ration. Energy is generally supplied from cereal grains and by-products, root crops, fats, sugar by-products or various fruits. Protein sources include oilseed meals, fish meals, animal by-products and various plants and nuts. Most of these sources contain some minerals and vitamins, and the remaining requirements need to be met by commercial mineral/vitamin supplements. The nutrient contents of the feedstuffs likely to be available for pig feeding in the tropics are given in Appendix B.

Major energy resources

Maize

Maize grain is probably the most extensively used energy source, certainly in Africa. It is very palatable and has a high energy value. Lysine and then tryptophan are the first two limiting amino acids when it is used for feeding pigs. If used at high concentrations in restricted diets, the oil content can cause soft fat in the carcass. Carcasses with soft fat are more difficult to store in an edible condition and are less acceptable to the consumer. Maize by-products (the part remaining after the grain has been processed for human food), such as bran, germ meal and hominy chop, are of higher fibre and lower energy content than the grain, but can also be fed to pigs.

Sorghum (Guinea corn)
Sorghum is often grown in drier regions as it tolerates drought better than maize. Trials in Zimbabwe indicate it has approximately 95 per cent of the feeding value of maize. Some varieties have a high tannin content, which lowers the palatability and protein digestibility of the grain.

Millets
If properly ground, millets make an excellent direct substitute for maize, and tend to produce firmer fat in the carcass. Bulrush and pearl millet are susceptible to infection with ergot fungus, which can cause agalactia (lack of milk). These varieties should not be fed to breeding sows.

Wheat
Although largely used for human food, wheat has a feeding value for pigs equal to maize if not ground too finely. Where wheat is milled for human use, the by-products (such as bran, middlings and weatings) are of considerably lower energy content but can be used.

Cassava (manioc, tapioca)
Production of cassava is widespread throughout Latin America and parts of Africa, mainly on small farms as a staple human food. It is high in energy but low in protein and this factor must be taken into consideration when cassava is used as a substitute for cereals in pig diets. Large amounts of cassava meal are exported from South East Asia to Europe for feeding pigs. The roots contain approximately 35 per cent dry matter and can be fed fresh, forming a particularly useful source of energy for dry sows. However, care must be taken since some varieties contain cyanogenetic glucosides, which can cause prussic acid poisoning. These varieties should be chopped, dried and ground to destroy any free cyanide. Research in Zimbabwe has shown that cassava can replace maize in pig rations with no loss of palatability, but extra protein must be added to the ration. Dried cassava leaves can also be used as an additional source of protein.

Sweet potatoes
These are often grown on small farms. In the fresh state, the tubers contain approximately 32 per cent dry matter and are best used as a feedstuff for sows. When dried and ground, the meal is similar in energy content to maize meal but very low in protein.

Potatoes
Surplus potatoes or those that are unfit for human consumption can be used to feed pigs. They contain about 22 per cent dry matter, high levels of energy (in the form of starch) and protein of very low biological value.

Potatoes must be cooked before they are fed to pigs because they contain a trypsin inhibitor, which reduces protein digestibility, and a toxin (solanin). Both substances are destroyed by cooking.

Yams

Yams are grown extensively for human food in West Africa and are not generally available for pig feed. They contain high levels of starch and, in the fresh state, they also contain a toxic alkaloid (dioscorine), and tannins, which give a bitter taste and reduce digestibility. Therefore they should be cooked before feeding to pigs.

Rice

Rice has about 75 per cent of the feeding value of maize for pigs and contains a relatively high fibre content. However, in South East Asia and Madagascar, the grain is the staple food crop and normally too valuable to give to pigs. Rice by-products, on the other hand, are often the sole source of animal feed in smallholder village production systems. These by-products, known as brans, millings and polishings, consist of the germs and outer layers of the grain. These can represent 8–15 per cent of the weight of the crop, and their use as pig feed partly explains the importance of pigs in these rice-growing areas. Rice that has been de-husked by an industrial process will produce a high quality by-product. However, traditional technologies do not completely separate the husk and bran, and the product has a higher fibre content and an excess of silica. 'Coarse' rice brans that include the germ are rich in polyunsaturated fats. This can lead to oily fat in the carcass if too much is given to pigs.

Molasses

This by-product of the sugar industry contains highly digestible sugars. It is highly palatable and can be used to improve the palatability of diets. Molasses can also be a good source of minerals and the B vitamins. However, it has a laxative effect if fed in large quantities and should generally be restricted to less than 10 per cent of the diet.

Citrus pulp

This is available as an energy source in citrus-producing areas. It has around 47 per cent of the feeding value of maize when used at low levels in the diet. Citrus pulp can cause diarrhoea if given in excess and it is recommended to include it at not more than 10 per cent of the ration.

Fats and oils

In the tropics, animal fat and vegetable oils are generally in great demand for humans and there is little available for pig feed. Nevertheless, they provide an excellent form of energy and confer additional benefits in terms of binding the dust in feed, providing essential fatty acids and a mild laxative action. If extraction facilities are not available, whole ground oilseeds can be used in pig feed. Full-fat soyabean, for example, is a good source of protein and energy. The raw soyabean contains anti-nutritional factors, such as trypsin inhibitors and the toxin, soyin, and must be heated to destroy them. In Zimbabwe, small processing plants have been developed to heat-treat whole soya and render it suitable for pig feed. In smallholder village systems, soyabean can be detoxified by boiling.

Protein sources

Soyabean meal

Extracted soyabean meal is one of the most widely used protein sources for pigs, and it possesses the highest biological value of the vegetable proteins. The first limiting amino acid in soyabean meal is methionine, which makes it an ideal combination with maize. Anti-nutritional factors in the raw soyabean are destroyed by the heat of the oil extraction process.

Groundnut meal

Groundnut meal is another widely used source of protein for pig feeding in the tropics. Although high in protein, the biological value is relatively low. It is particularly deficient in lysine and methionine. Owing to the susceptibility of groundnuts to infection with moulds, the meal often contains aflatoxins, which seriously depress the growth rate of pigs. In some countries (e.g. Madagascar) village oil mills produce a meal product with a high oil and energy value. This serves as a very convenient source of protein for smallholder systems of pig production.

Cottonseed meal

Cottonseed meal is deficient in lysine and tryptophan and is not an ideal protein supplement for pigs. Raw cottonseed contains a toxin, gossypol, which can prove fatal to pigs. Although gossypol is largely inactivated by the oil extraction process, it is best to feed no more than 10 per cent cottonseed meal in pig rations. If necessary, gossypol can be neutralised by the addition of one per cent iron sulphate. Gossypol-free varieties of cotton are currently being distributed to farmers in Central and West Africa, and this is likely to increase the use of cottonseed meal in pig diets in the future.

Sunflower meal

Sunflowers are grown fairly extensively by small-scale farmers. The meal is high in fibre (11–13 per cent), and the protein is very deficient in lysine, and is therefore only recommended for inclusion in pig diets at low levels.

Sesame meal
Extracted sesame meal contains 40–45 per cent protein, which is a poor source of lysine but has high levels of methionine and tryptophan. It is therefore useful for blending with other proteins.

Coconut oil meal
Coconut oil meal contains about 20 per cent protein and is high in fibre. Protein digestibility is often low, and it is deficient in lysine, factors that can depress growth rates unless the meal is used at relatively low inclusion rates. In some situations, as in the Pacific islands, coconut meal may be the only locally available by-product containing protein. It may then be used extensively in pig diets.

Palm kernel meal
An increasing area is being devoted to oil-palm cultivation in the tropics, therefore the meal which remains after oil extraction from the seeds is likely to become more plentiful. However, because the meal is high in fibre, relatively unpalatable and contains protein of only medium biological value, it should be used only in small amounts.

Safflower meal
Owing to the high proportion of hulls in the meal, the high indigestible fibre content severely limits the use of this meal in pig diets. The feeding value is estimated to be about 60 per cent that of soyabean meal.

Beans and peas
There are various species of peas and beans grown throughout the tropics that can be fed to pigs, e.g. kidney beans, haricot beans, jackbeans, cowpeas, chickpeas and pigeon peas. They all have a similar protein content (18–25 per cent), and are generally deficient in methionine and sometimes tryptophan. Anti-nutritional factors are common in most species and need to be deactivated by cooking.

Rubber tree seeds
In low-input pig production systems in South East Asia, pigs are left to scavenge for rubber tree seeds when available. Although the seeds may contain toxic cyanic acid when fresh, this is eliminated with age, and the pigs learn to harvest only 'mature' seeds. The seeds contain 17 per cent protein and 18 per cent edible oil.

Fish meals
These are excellent sources of protein for pigs, with a high biological value and high lysine content. They are palatable and contain a good source of calcium and phosphorus. However, their use in the tropics has declined in recent times due to lack of availability and high cost.

Meat and meat-and-bone meals

Meals produced from abattoir waste products tend to vary considerably in quality and are not generally in large supply. Unless stabilised by an antioxidant, they also have the problem of the fat going rancid and reducing palatability. Nevertheless, if of good quality, they can make an excellent supplement for cereal-based diets, and are generally included in pig diets at between 5 and 10 per cent. Care must be taken that they are processed properly and do not transmit diseases such as anthrax and foot and mouth disease. Because of the risk of disease transmission, particularly of Bovine Spongiform Encephalopathy (BSE), the use of abattoir by-products in livestock feeds is now banned in large parts of the developed world.

Blood meals

These are of relatively low palatability, which limits their use in pig diets. However, they can be a valuable supplement because they are particularly rich in lysine, but very often the protein and amino acids are damaged in the drying process.

Feather meal

Feather meal, a by-product of the poultry industry, has a very high protein content (85 per cent). It must be hydrolysed to promote availability of the protein from keratin. At a low inclusion rate, feather meal can be useful as it is rich in cystine, threonine and arginine.

Milk products

Milk is normally too much in demand for human consumption to be used in pig feeds. However, its digestibility and protein biological value are very high. Skimmed milk, which remains after the removal of butter fat from whole milk, can be fed in fluid form or dried to give a powder containing 33 per cent protein. It is particularly useful to include skimmed milk in creep rations for piglets as it improves the palatability and digestibility of other feeds. Whey, which is a by-product of the cheese-making process, contains only about 5 per cent dry matter, but can be dried to yield a powder of 15 per cent protein. If milk products are fed in the fluid form, they should be fed either consistently fresh or consistently sour because the pig's gut microflora will adapt accordingly.

Some other bulky feeds

Bananas

Ripe bananas are very palatable to the pig, and studies have shown that the growing pig will consume up to 8 kg per day. Bananas contain just 1 per cent protein, but can provide a useful supplementary source of energy in the form of sugars.

Brewing by-products

These consist of either wet or dried brewers' grains from barley, sorghum or millets. In the process of brewing, the soluble carbohydrates are utilised to produce alcohol so the residue is richer in protein and fibre than the original grain. Although the high fibre content (15 per cent), low protein and great variability limit their feed value for pigs, they tend to be widely used in small-scale systems of pig production, especially in Africa. Residues from traditional beer brewing systems are also often available to pigs kept on scavenging systems of production.

Water hyacinth

Water hyacinth, a troublesome weed of waterways throughout large parts of the tropics, is used as a feed for pigs in South East Asia. The plants are chopped, sometimes mixed with other vegetable waste, and boiled slowly for a few hours until they turn into a paste. This is supplemented with oil cake meal, rice bran and sometimes maize or salt. A common ration formula contains 40 kg water hyacinth, 15 kg rice bran, 2.5 kg fish meal and 5 kg coconut meal.

Leaf protein

High levels of fibre limit the amount of leaf protein that can be included in pig diets, however, green leaves such as lucerne (alfalfa), other legumes, cassava, sweet potato vines and grass can be dried and ground or fed fresh to pigs. The leaves should be as young as possible to minimise fibre content.

Swill

Swill or kitchen waste from institutions, hotels, restaurants, etc. can be a major source of feed for pigs, especially when pigs are reared in close proximity to a town or city. The nutrient content of the swill can be very variable according to the source. It can also transmit infectious diseases (e.g. swine fever and foot and mouth disease), so all swill should be boiled before feeding to pigs. Swill often contains broken glass, fish bones and pieces of metal, which can be harmful to pigs. The best system of feeding swill is to use the 'Lehmann system', which was developed for bulky feeds. On this system, the pigs receive a set amount of balanced concentrate each day and then have *ad lib* access to the swill (they can eat as much as they like).

Pumpkins and melons

Pumpkins and melons grow well in the tropics and are often fed to pigs on small-scale systems. The nutritive value is low and they contain virtually no protein, but they can provide some energy and succulence, particularly on scavenger systems of production.

Genetically modified feeds

In spite of the consumer resistance to the genetic modification of crops used for human food, there are exciting possibilities for the use of genetically modified pig feeds to improve the efficiency of feed utilisation and the health of pigs. Taking soyabeans as an example, there are some interesting developments:

- A variety producing large amounts of the phytase enzyme. This will increase the availability of phosphorus in the soyabean meal, reduce the amount of phosphorus excreted into the environment and improve protein digestibility.
- High lysine and high methionine soyabean meals, which improve the biological value and nutritional quality of the soyabean protein.
- Varieties with lower then normal levels of the insoluble carbohydrate, stachyose. This increases the energy content of the meal and reduces manure solids.
- Varieties containing high levels of oligofructans. These act to selectively increase the population of beneficial bacteria in the intestines and keep the pig healthier.

6 Housing

Pig productivity and comfort

If pigs are to achieve maximum productivity, they need to be kept in a thermally neutral environment where the environmental temperature around the pig remains between the pig's lower (LCT) and upper critical temperature (UCT) (see Chapter 2). The pig's metabolic heat production is then at a minimum and it is neither using feed energy to keep warm, nor reducing feed intake to keep cool. Other considerations for pig comfort and well being, in addition to temperature, include:

- protection from other climatic extremes such as direct sun, wind and rain;
- provision of dry conditions that are hygienic and do not predispose the pig to disease;
- allowing, as far as possible, for inherent behaviour patterns and minimising the effect of social dominance;
- provision of accessible feed and clean water;
- good stockmanship;
- effective disposal of effluent.

General design considerations

Any buildings, whether simple or complex, cost money to build and maintain. It is therefore imperative that careful thought is given to the design, so that the investment is justified by improved productivity. Moreover, considerations that affect design of houses in the tropics can be very different from those in more temperate parts of the world.

Under tropical conditions, the main consideration is usually to ameliorate the effects of excess heat. At the same time, it is important to minimise temperature variations, keeping as close as possible to the pig's zone of thermal neutrality. This often involves keeping pigs cool by day and warm by night, particularly at high altitudes where diurnal fluctuations in temperature can be as great as 25°C.

Site

Although the choice of a site may be limited, it is often possible to take advantage of the natural prevailing winds to increase ventilation and airflow. Buildings should face north–south to avoid excessive penetration of sunlight into pens, preventing problems of sunburn and heat stress. At the same time, the site should preferably not be too exposed to cold winds. Ideally, pig buildings should be on a slight slope to facilitate drainage and disposal of effluent. The slope also makes it easier to design a 'pig-flow', with the farrowing accommodation at the top of the slope and the fattening pens at the bottom. This is likely to improve management and prevent cross-infection from effluent between categories of pigs, an important concept in preventing disease in the most susceptible groups, especially piglets.

Other considerations include proximity to a good water supply, ensuring that effluent cannot contaminate water supplies from boreholes or wells, and access to a good road (especially for more intensive piggeries) to allow for transport of feed in and pigs out. Finally, allowance should be made for adequate space between buildings and for possible expansion in the future.

Floors

Even in low-cost buildings, hard concrete or similar floors are recommended. These prevent the pigs 'rooting' and digging up the ground and, most importantly, facilitate cleaning and reduce disease and parasite problems. The floor should provide insulation against cold and damp. Many insulating materials can be used, including those that create an under floor airspace, such as old bottles, air bricks, piping, etc., but stone or gravel is most common. Fig 35 illustrates a suitable piggery floor. The nature of the floor surface is important. If the surface is too smooth and slippery, injuries will occur. If it is too rough or abrasive, feet and udders may be damaged.

Roofing

Roofs need to provide shade and protection from the weather. They can be made from local materials such as grass, reeds, leaves, etc. (Fig 36) or from manufactured roofing material such as tin or asbestos. A thatched grass roof is appropriate for hot conditions because it is well-insulated. However, if not protected, it rapidly becomes a breeding ground for rats

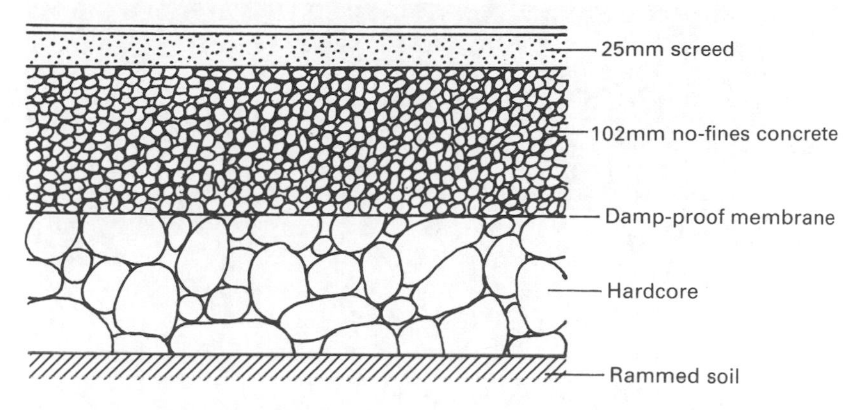

Fig 35 *A suitable piggery floor*

Fig 36 *Simple low-cost housing made from local materials*

and other pests. In hot climates, asbestos or tin roofs should be insulated with a layer of grass, insulating board or equivalent, or be sufficiently high that there is room for plenty of air movement. On an apex roof, there should be a raised ridge or other arrangement to promote ventilation and allow hot air to escape (Fig 37).

Fig 37 *Raised roof ridge allows hot air to escape*

Walls

External walls should be strong enough to contain and protect the pigs. They can be built from a variety of materials, ranging from mud and/or sticks (Fig 36) through to bricks and cement. In enclosed buildings, it is important to provide suitable flaps, windows and doors that can be opened to allow maximum ventilation during hot periods, and closed to retain warmth at colder times (Fig 38).

Fig 38 *Door flaps maximise ventilation when hot and retain warmth in cool temperatures*

Space

Overcrowding is a common cause of depressed performance and productivity in pigs. Recommended space requirements for various categories of pigs are shown in Table 11.

Table 11 Recommended space requirements for pigs

	Area
Boars	9 m^2
Dry sows	
Stalls	2 m long x 0.64 m wide
Cubicles	As for stalls plus similar exercise area
Yards	3–4 m^2 per sow
Farrowing accommodation	
Pen, including crate	6.2 m^2
Fattening/follow-on (including creep area)	10 m^2
Weaners	
Cages (per pig)	0.2 m^2 lying area + 0.2 m^2 slatted area
Yards (per pig)	0.7–0.9 m^2
Porkers	
Pen, including dunging area	0.73 m^2 per pig
Baconers	
Pen, including dunging area	0.93 m^2 per pig
Heavy pigs	1.10 m^2 per pig
Trough space (per pig)	
Fatteners	0.2–0.3 m^2
Maiden gilts, sows	0.35 m^2

Boar housing

Protection of boars against temperature extremes is very important, as high temperatures can rapidly lower fertility of semen and reduce libido. Boar accommodation must be strongly constructed, particularly to prevent boars escaping and fighting (Fig 39). Wire reinforcement can be used for walls, and bolts and hinges for doors must be robust.

Sows should not be served by the boar in the boar pen – a simple service area should be constructed outside (Fig 40). This will be cooler in the

Fig 39 *A strongly constructed boar pen*

early morning and evening, it provides a good foothold to the pigs on the grass or soil, and it avoids any right-angle corners, which can interfere with service procedure.

Dry sow housing

Yards

Yards are the simplest form of dry sow housing, and consist of a yard or large pen shared by several sows (Fig 41). Groups of sows should not be too large to avoid bullying and fighting. Ample shaded area and trough space must be provided. Ideally, sows should have access to separate feeders so that feed intake for individual sows can be controlled. This system is widely used by small-scale and less intensive producers. It is the best system for gilts as it allows them to interact and stimulates oestrous behaviour.

Fig 40 *A cool area in the piggery used as a service pen*

Fig 41 *Dry sows can be housed in groups*

Tethers

The system of tethering sows on long leads, which is commonly seen throughout the tropics, can be used within yards or paddocks as a means of restricting sows to a given area.

Sow stalls

In sow stalls, each sow is individually penned and fed. The main advantages of individual stalls are that bullying and fighting are eliminated, cleaning-out, feeding and management are facilitated, sows can be fed according to individual requirements and the sow does not use up energy in walking. The major disadvantage is that it can lead to lameness and lack of muscle tone. Sow stalls must be designed so that the pigs are protected from the weather on all sides of the building, since they cannot huddle together to conserve body warmth in cold conditions.

Cubicle housing

In this type of accommodation (Fig 42), sows have a communal dunging and exercise area and each sow has a stall for lying and feeding. The dunging passage can also be used for the boar to make his daily rounds for heat detection, and the design normally incorporates boars' quarters. Cubicles tend to combine the advantages of sow stalls (individual feeding and attention) and yards (exercise).

Electronic sow feeders

The advent of computer-controlled electronic sow feeding systems in developed countries combines the advantages of yards and individual sow feeding. Sows housed communally can gain access to individual feeders by means of transponders installed around their necks or in their ears. However, this level of sophistication is unlikely to be in general use in developing countries for some time.

Farrowing and rearing accommodation

This is probably the most important building in the piggery, and needs to provide suitable conditions for the birth of piglets and getting them off to a good start in life. The need for specialist design in farrowing pens has increased as sows have been bred with larger body sizes, bigger litters and reduced mothering ability. Thus, while the small indigenous sow will make a nest, farrow and rear her piglets very efficiently without any help, the average exotic sow needs certain provisions to maximise piglet survival. The essential features of farrowing house design are:

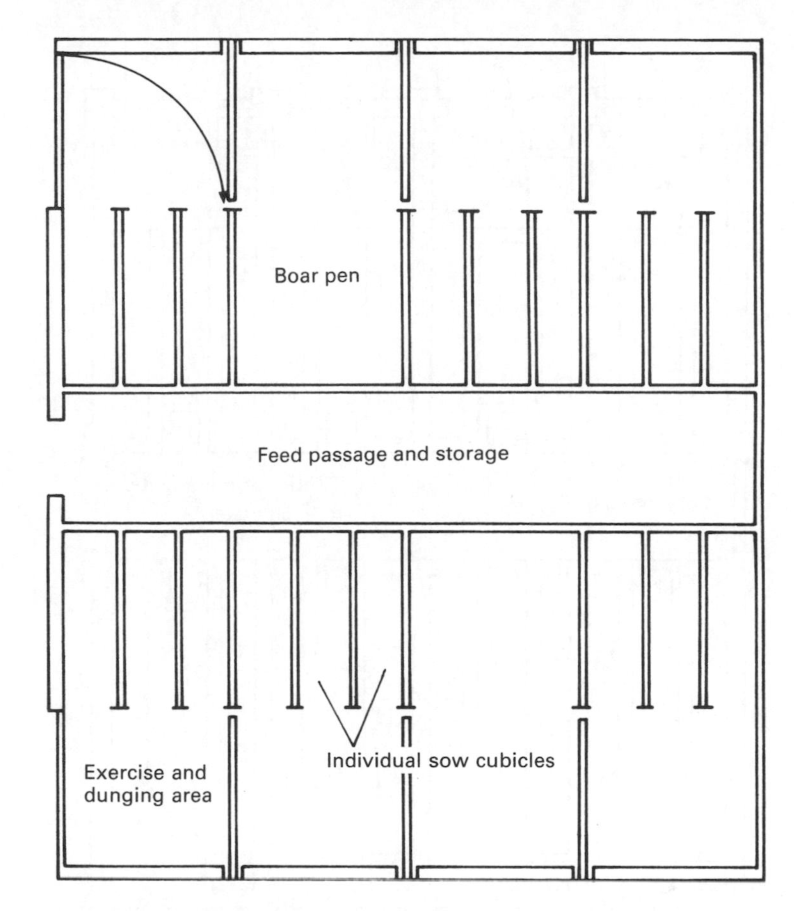

Fig 42 *Plan of cubicle house*

- protection of the piglets from being crushed by the sow;
- provision of higher temperatures for the piglets than for the sow. If the whole house is the ideal temperature for the piglets, this will reduce the sow's appetite;
- easy access for cleaning so that high levels of hygiene can be maintained.

Farrowing crates

These are generally incorporated into specialist farrowing houses (Fig 43). The sow is restrained in the crate while the piglets have access to a small surrounding pen, which includes a creep box, creep feeding facilities and their own water supply. The design and dimensions of the crate are important to ensure both sow comfort and farrowing efficiency (Figs 44 and 45). Crates have several advantages, as they:

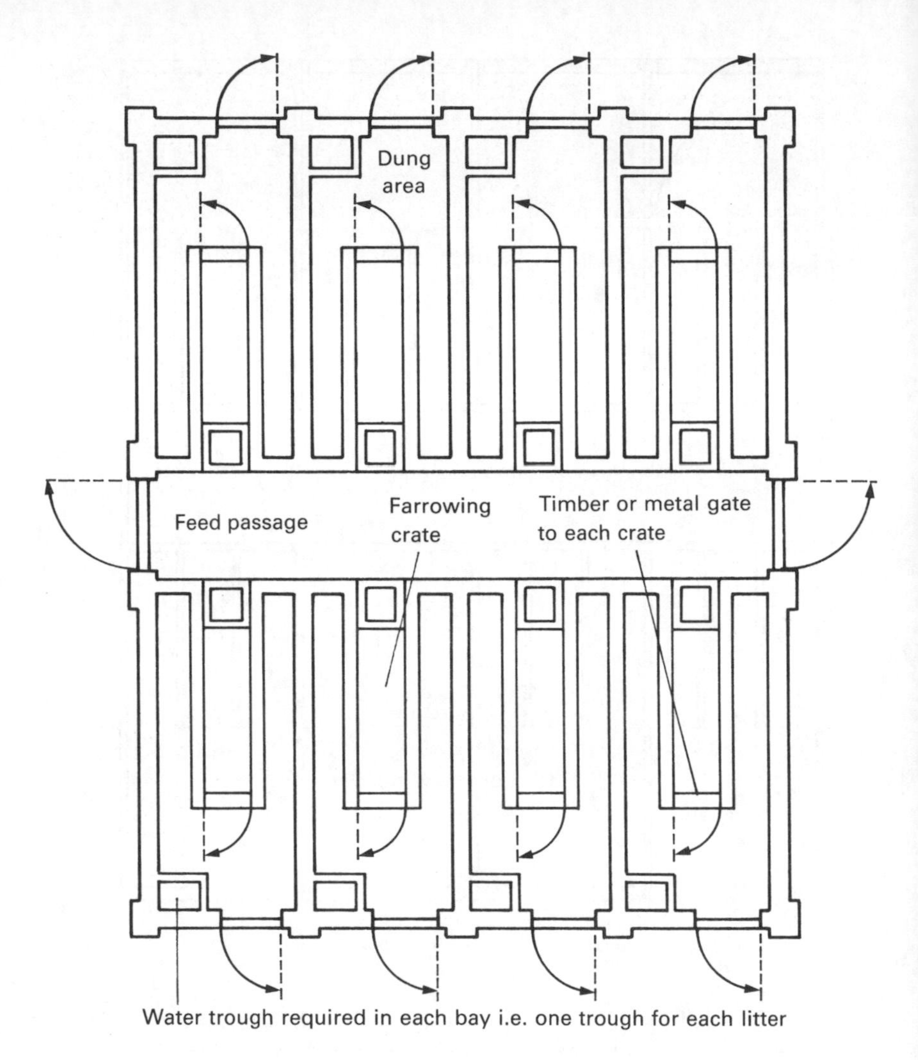

Fig 43 *Plan of farrowing house*

- keep piglets close to the sow, but prevent her from lying down quickly and crushing them;
- allow the stockman to handle the sow and her piglets without risk of injury;
- allow piglets free access to creep feed and water.

Even under tropical conditions, a separate creep area for the piglets, which is warmer than the ambient temperature, is generally an advantage, especially at night. This is because the optimum environmental temperature for the sow is between 16 and 18°C, whereas that of the newborn pig

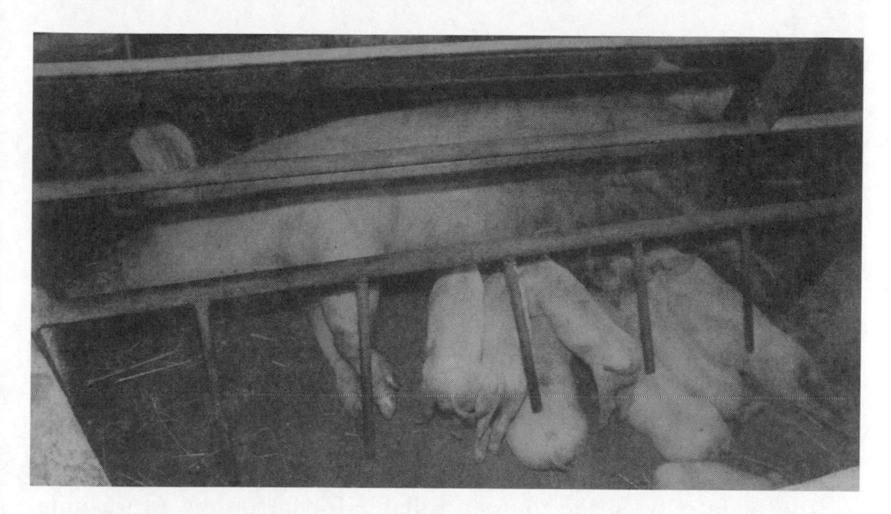

Fig 44 *Specialised farrowing pen containing a farrowing crate*

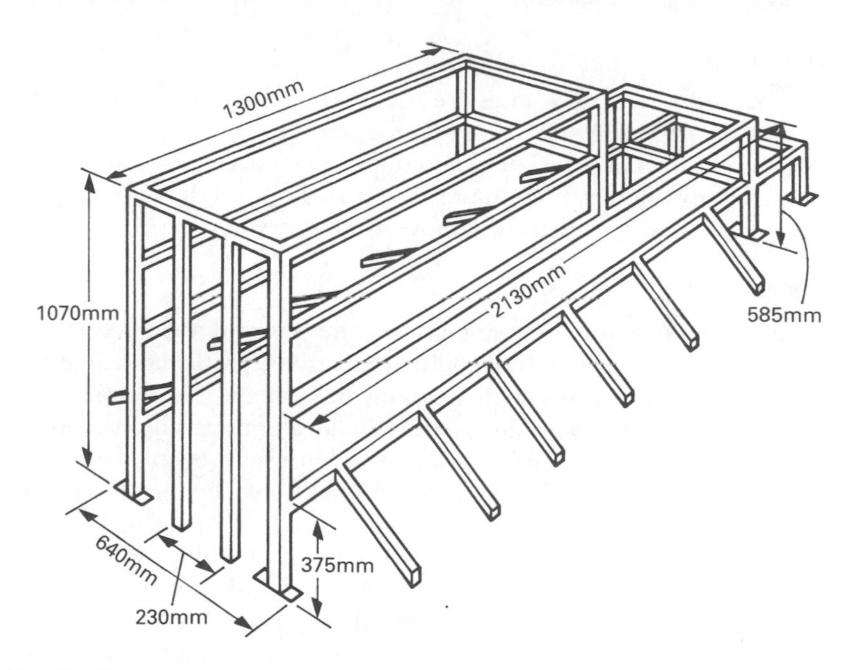

Fig 45 *Farrowing crate showing dimensions*

is 33–35°C. A simple, enclosed creep box (Fig 44) allows the piglets to generate their own warmth by huddling together. If electricity or paraffin is available, a light or simple heater can be provided in the creep box. This not only provides extra warmth, but also attracts the piglets into the box and away from the danger of being crushed by the sow.

If separate arrangements are not made for the piglets and the whole farrowing room is warm, sow productivity will be reduced. American work has shown that for every 1ºC rise in temperature from 25–30ºC, daily feed intake by the sow declines by 400 g. This is important because sows, even when kept under cool conditions, have difficulty in consuming enough feed each day to maintain bodyweight and produce enough milk to meet the needs of their piglets.

Multi-purpose pig pens

These are more appropriate in the tropics and the developing world, as they are cheaper and more flexible than farrowing crates. Removable structures, such as creep barriers and farrowing rails, provide protection for the piglets and make the pen suitable for farrowing. At weaning, these are removed, leaving a fattening pen in which the weaners can be fattened through to slaughter (Fig 46).

Follow-on pens

If specialised farrowing houses are used, sows can normally be moved into cheaper housing after 10–14 days when the danger of sow-related piglet deaths has passed. Piglets then require a separate creep area, similar to that provided by a multi-purpose pen. This has the advantage of allowing the sow to exercise and move around freely.

Weaner cages

The combined trauma of weaning from the sow and a change in diet makes the young pig very susceptible to diseases, particularly digestive diseases, which can cause high mortality in weaners. The weaner cage was originally designed in Europe with the idea of providing conditions to overcome such problems. Weaner cages have since been adapted for hotter climates and consist of a covered solid floor sleeping and eating area and a dunging area floored by either wooden slats or metal mesh (Figs 47 and 48). During cold periods, pigs can huddle and generate heat inside the covered section. Ventilation is provided by the centrally-hinged internal roof (Fig 48). In hot weather, pigs can keep cool by lying on the meshed floors, and are protected from the sun by an umbrella roof over all the cages. As dung and urine falls through the wire mesh or slats, this can be cleaned from below and there is therefore no need for stockmen to enter the cage with contaminated boots, brooms or shovels. Pigs normally remain in the cages for three to four weeks before being transferred to fattening accommodation. The feed hoppers can be moved to allow additional space as the weaners grow.

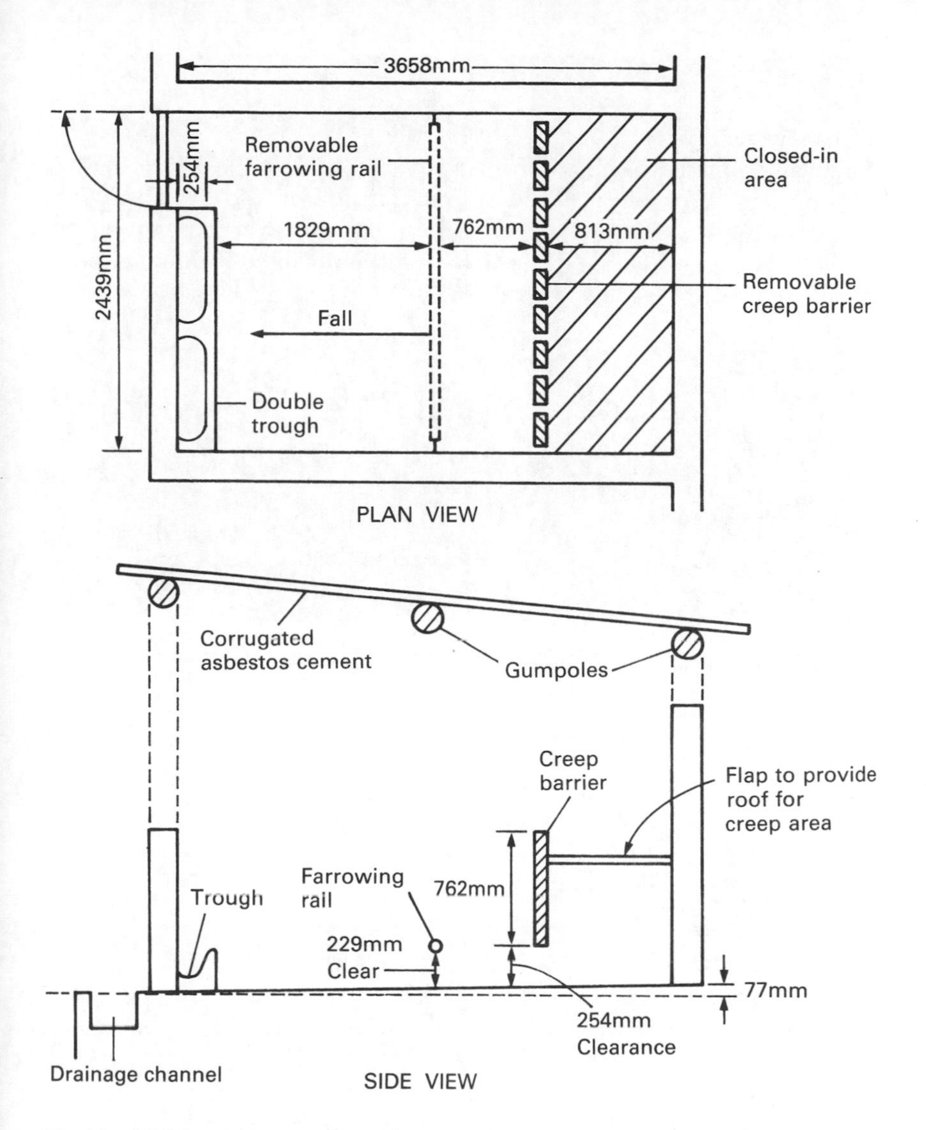

Fig 46 *Multi-purpose pen – farrowing structures are removable and the pen can be used for fatteners or other stock*

Weaner pools

The traditional system of housing weaners is to take litters of similar ages and move them into large pens holding up to 50 weaners. After three or four weeks, pigs are batched (sorted) into groups of equal sizes for transfer into growing/fattening pens. Ample watering and trough space must be provided, and some form of bedding is preferable. A sheltered kennel

Fig 47 *The front of a weaner cage*

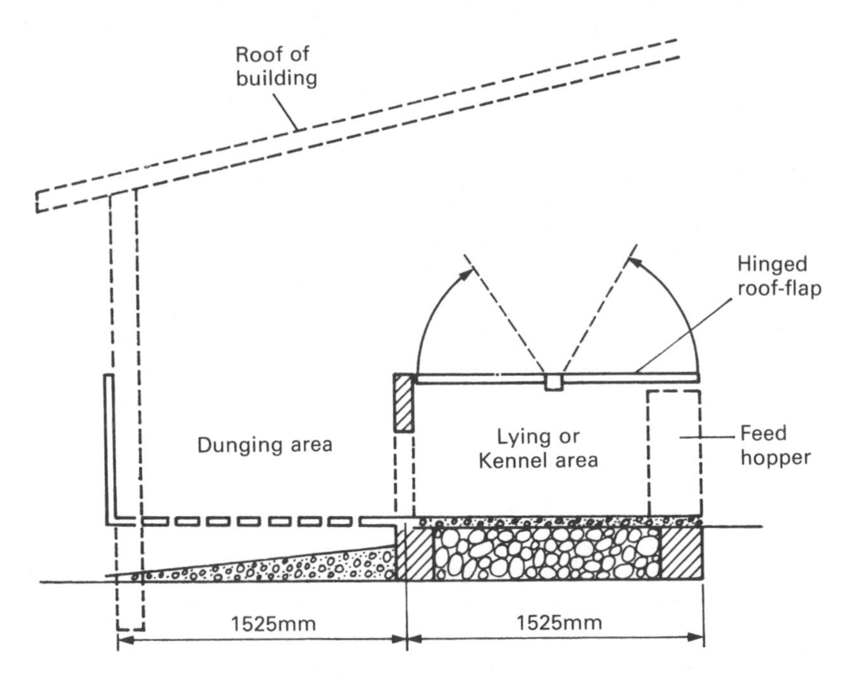

Fig 48 *Side plan of a weaner cage*

area, which can be insulated by a ceiling of hay bales or wood, can be provided for added warmth.

Specialist weaner accommodation

With the increasing realisation of the importance of temperature on newly-weaned pig performance, there has been a trend among more intensive producers to provide specialist weaner accommodation, which generally involves smaller pens housing 10–20 pigs. These provide greater control over the pig's ambient temperature, and the kennels may be heated at night or during cold periods. Temperature can be adjusted as the pig grows to prepare it gradually to the temperature of the grower accommodation.

Growing/fattening accommodation

The basic needs for good fattening pens are relatively simple, namely, a dry area and a demarcated dunging area (Fig 49). The building should provide shade, some protection and adequate ventilation. In hot climates, high roofs are an effective way to improve ventilation and keep the pigs cool (Fig 50). Solid walls are not required between pens, because they will reduce ventilation and airflow within the building. Pens designed to hold 8–10 pigs through to slaughter are the ideal size.

Extensive systems

Extensive systems (see Chapter 3) are particularly appropriate for sows. Sows are run in paddocks and have access to arks or huts in which to farrow. In trials in Zimbabwe, sows were allowed a choice of different designs of arks at farrowing time, and it was found that they preferred a design similar to those found in the UK (Fig 51). The major difference is that under tropical conditions, the roof should be insulated with a 5 cm layer of grass or similar material. Arks can be constructed from cheaper materials, but it is difficult to make them sufficiently robust to avoid destruction by the sows. Ample deep shade and wallows should also be provided for sows run under this system. As mentioned previously, tethers can be used as a means of restricting sows within a paddock. They can be rotated around a given area of pasture or other forage.

In Africa, a major extra cost in extensive systems is the protection of pigs against African swine fever. This requires two robust boundary fences, about one metre apart, which prevent physical contact between domestic pigs and bush pigs or warthogs, which are carriers of the disease.

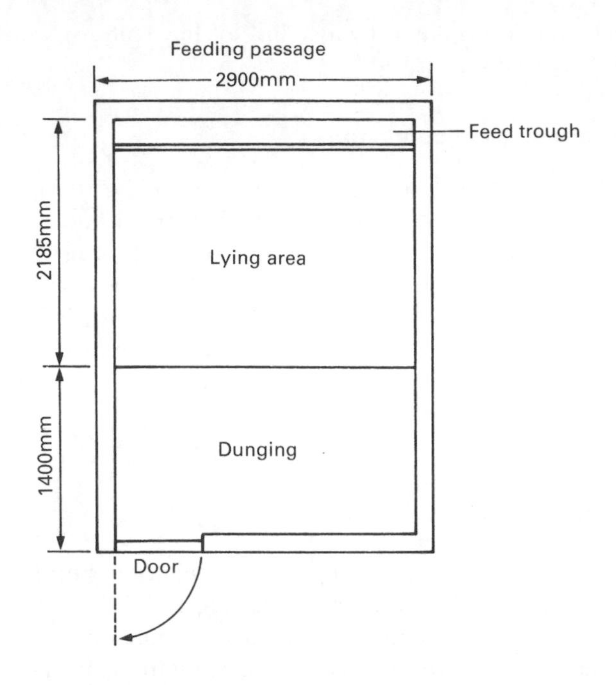

Fig 49 *Floor plan of a growing/fattening pen designed to hold 8–10 fatteners*

Fig 50 *Growing/fattening building with a high roof to improve ventilation*

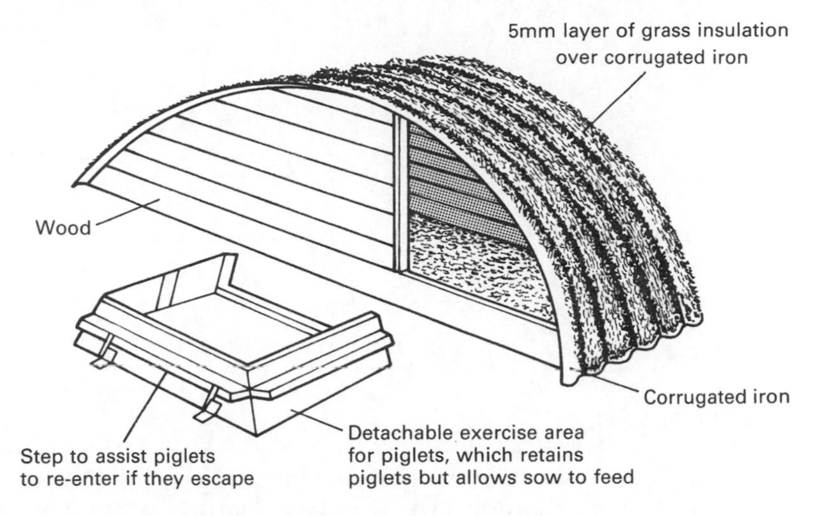

Fig 51 *A farrowing ark for use under extensive systems*

Welfare considerations

The last decade has seen an upsurge in consumer concern for the welfare of domestic animals. In some developed countries, lobbying has led to prohibition of pig housing systems that are seen to restrict movement, (e.g. dry sow stalls, farrowing crates and sows on short tethers), lead to overcrowding, prevent social interactions or provide insufficient comfort (e.g. lack of bedding). Although these concerns and prohibitions have not yet affected the developing world to any extent, they will clearly be a consideration for future planning and design of pig housing. Further information can be found in *Livestock Behaviour, Management and Welfare* in this series.

7 Health

Disease prevention

Disease in a pig herd can have an enormous impact in terms of the cost of control and reduced productivity. The first priority, therefore, is to prevent the occurrence of disease. Thus, many of the management procedures discussed in Chapter 8 aim to prevent disease or mitigate the effects of diseases that cannot be prevented. Skilled management, well-designed housing and adequate nutrition are all important features of a disease prevention strategy.

Nevertheless, a basic knowledge of the main diseases that may affect a pig herd is necessary so that a producer can diagnose the condition and implement control measures as quickly as possible. This is particularly useful in tropical conditions, where veterinary services may not always be available. The major disease problems are parasites, infectious disease and a few non-specific diseases. Conditions associated with nutritional deficiency have been dealt with in Chapter 5. More information can be found in the two volumes of *Animal Health* in this series.

Parasites

Parasites are organisms that live on and obtain feed from the body of another, known as the host. They may live on the exterior of the pig, when they are known as external parasites, or within the internal tissues and organs, when they are referred to as internal parasites. The presence of parasites will not usually be fatal for the host, except in the case of massive infestations or when the host is subject to additional stresses.

External parasites

These irritate the skin and often lead to wounds and increased susceptibility to other infections. The most common external parasites are mange-mites, ticks, lice, fleas and flies.

Mange-mites
Mites, which are scarcely visible to the naked eye, spend their entire lifecycle under the skin of the pig, but they can survive off the host for as long as eight days. The most common species is *Sarcoptes scabiei,* which causes sarcoptic mange. The first sign of infection is a crusty, dry-looking skin around the eyes, ears and snout. The mites then spread and multiply over the body, and their burrowing causes the skin to become inflamed. The pig will constantly rub itself and its performance will be depressed. Control is best achieved by regular treatment, either by dipping or spraying the animal with an anti-mange medication and by spraying the pen. Chronically infected animals should be culled. Systemic drugs have recently become available and are very effective against the mite.

Ticks
Ticks are only a problem in scavenging or more extensive systems of pig production. There are a number of different species, but all suck blood and can transmit serious diseases (e.g. Babesiosis or redwater). They generally require more than one host to complete their lifecycle. Ticks are easily controlled by spraying or dipping with suitable acaricides.

Lice and fleas
Lice and fleas can become a problem in dirty and unhygienic conditions. They live on the skin surface, suck blood and cause irritation. Spraying pigs and pig pens with suitable insecticides is an effective means of control. In the case of lice, particular attention should be paid to the ears.

Flies
Flies have a major nuisance value around pigs as they cause annoyance, aggravation and considerable stress. They can also bite and carry infectious diseases. They are attracted to any fresh abrasion or wound on the animal, and are a particular problem in housed pigs, which cannot get away from the irritation. As tropical climates particularly favour the build-up of flies, efforts to control fly populations are essential. Control measures should involve spraying of insecticides on suitable fly-breeding areas, e.g. manure heaps, refuse areas and ponds, pig buildings and the pigs themselves. Baits that attract the flies and are poisonous to them (but not to the pigs) can also be effective. New compounds can be added to pig feeds and pass out in the dung where they interrupt larval development. (See section on trypanosomiasis for information on the tsetse fly.)

Internal parasites

Roundworms
These are a particular hazard when pigs are free ranging or kept on non-solid floors. The large roundworm (*Ascaris lumbricoides*) is very common

and can cause a lot of damage in pig herds. Adults live in the small intestine and can grow up to 300 mm long and 6 mm thick (Fig 52). The female is capable of laying thousands of eggs per day, which pass out in the dung and become infective when ingested by other pigs 21 days later. The eggs are extremely resistant to the effects of the environment and can remain infective in the soil for many years. As part of the life cycle, eggs hatch out in the pig after ingestion and the larvae migrate through the liver and lung (Fig 53). Irritation in the lungs causes coughing and ill thrift, particularly in younger pigs. Liver damage, which appears as 'milk spot liver' (Fig 54), also occurs and renders the liver liable for condemnation at slaughter. Moreover, if infection is heavy, the adult worms can partly obstruct the small intestine, causing weakness and weight loss.

Fig 52 *Section of the small intestine showing severe infestation with roundworms*

Another harmful roundworm is the whipworm (*Trichuris suis*). The adults grow to around 35 cm long and cause considerable damage to the wall of the large intestine, causing diarrhoea and weight loss. The nodular worm (*Oesophagostomum spp.*) also lives in the large intestine. It burrows into the intestinal wall and forms nodules, causing diarrhoea (sometimes bloody) and anaemia. The kidney worm (*Stephanurus dentatus*) lives in the kidney and its eggs are excreted via the urine. When ingested, the larvae migrate through the liver to the kidney resulting in tissue damage. The kidney worm is a major handicap for free ranging pig systems in Madagascar and the Pacific Islands, and is often the main reason why pigs are

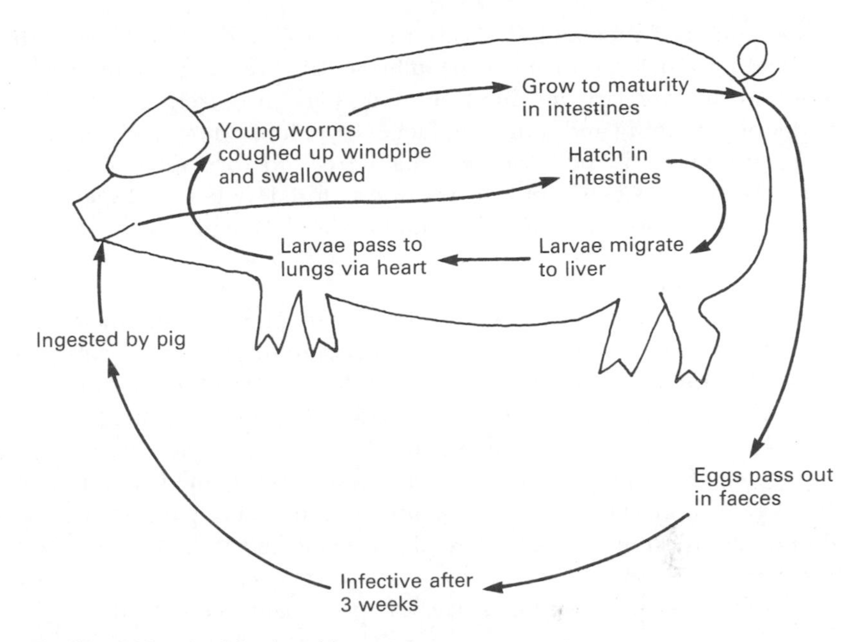

Fig 53 *Life cycle of the Ascarid worm*

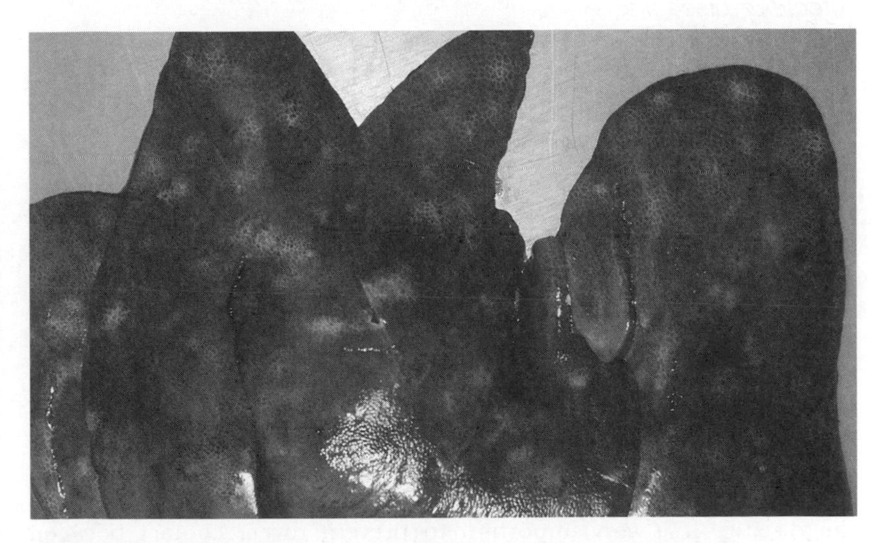

Fig 54 *Liver damage (milk spots) caused by larvae of* Ascaris lunbricoides

penned. Lungworms (*Metastrongylus spp.*) can also be a problem in free range pigs; infection occurs when pigs eat earthworms, which are the intermediate hosts. Lungworms cause irritation and coughing and predispose the lung to secondary pneumonia.

Contaminated feed and water are the usual sources of infection with internal parasites. Control can be achieved by breaking the life cycle, which means regularly moving free range pigs on to fresh ground and frequently cleaning and removing faeces from housed pigs. At the same time, unless there is good evidence that there is no worm infection in the herd, breeding pigs should be routinely dosed with broad-spectrum anthelmintics and young stock should be dosed soon after weaning.

Tapeworms

The common tapeworm is *Taenia solium*. The pig is its intermediate host and the adult worm lives in man. Pigs become infected by picking up eggs from human faeces and the larvae then encyst in the pig's muscle, particularly in the region of the heart and lungs. If the pig meat is then eaten by man, the larvae hatch out and the cycle is completed. As a consequence, affected carcasses (measly pork) are condemned at slaughter. By preventing pigs having access to human faeces, the parasites can be eliminated. In some countries, live pigs are checked at the market place by trusted experts for the presence of tapeworm cysts in the tongue. The result of the examination influences the price paid to the producer.

Infectious diseases

The following diseases are notifiable (i.e. their occurrence must be reported to the national authorities by law) in most countries.

African swine fever

This is a highly contagious viral disease and in the acute form can cause 100 per cent mortality. Typical symptoms are loss of appetite, huddling together, small purplish blotches on the skin, lack of coordination and laboured breathing. Although the disease was first reported in 1921, it has re-emerged during the past decade and now poses a serious threat throughout continental Africa. It has spread to many previously unaffected countries with devastating effects. For example, Benin had its first case in 1997 and subsequently lost half its national herd, and Côte d'Ivoire lost 25 percent of its pig population in 1996.

The virus is carried by bush pigs and warthogs, but they are immune to the disease. It is very important to prevent direct contact between domestic pigs and wild species. This can be achieved by double penning and by controlling animal movements. The disease can also be transmitted by a soft tick (*Ornithodoros moubata*), which infests the warthog. Otherwise infection occurs by contact with sick pigs, or through contaminated feed or water. There is no effective vaccine or treatment and infected pigs should be isolated. Recent research has suggested that some indigenous pig breeds in Africa may possess a genetic resistance to the swine fever virus.

Foot and mouth disease

Regarded as the most contagious of all known viral diseases, infection causes blisters on the feet, snout, udder and in the mouth and throat. It is very painful to the pig, which cannot eat and may have to be destroyed. This disease affects all domestic meat and/or milk animals except the chicken. Foot and mouth disease is endemic in parts of Africa and the virus is carried by the buffalo. Infection can occur by feeding infected bones or cooked meat, but can spread from animal to animal and from herd to herd by direct contact, by air, on the wheels of vehicles, etc. There is no cure. If an outbreak occurs in an adjacent area, pigs can be vaccinated, but as there are many different strains of the virus it is important to ensure that an appropriate vaccine is administered.

Anthrax

This is an acute and frequently fatal bacterial disease, which can also be fatal to humans. There are two main symptoms: swelling in the back region (which causes difficulty in breathing) and sudden death with blood oozing from the body orifices. If the disease is suspected, the carcass should not be opened as this would allow infective spores to escape. Infected carcasses should be buried at sufficient depth to prevent any transmission of the spores. Pigs are infected by contact with anthrax carcasses or by spores in contaminated feed or pasture. There is an effective vaccine.

Rabies

Pigs with rabies show intense excitement, bite other pigs, salivate, become uncoordinated and die within two or three days. The disease is also fatal to humans. Infected animals should be shot through the heart, as brain damage can result in the release of the virus. Pigs are infected by contact with the saliva of infected animals, particularly dogs and jackals. Such contact should be prevented.

Trypanosomiasis

Although trypanosomiasis is mainly restricted to the tsetse fly areas of Africa, pigs are nevertheless susceptible to the disease. It is caused by a protozoan, *Trypanosoma vivax* or *T. congolense*, which is transmitted by the tsetse fly. Infection will cause debilitation, anaemia and ultimately death. In some cases, the acute form of the disease will cause sporadic sudden death without symptoms. There are, however, effective drugs against the disease but these tend to be costly. Another form of trypanosomiasis occurs in Asia and Tropical America and appears to be on the increase. It is caused by *Trypanosoma evansi*, and is transmitted by species of *Tabanus*, a biting fly.

Other (non-notifiable) infectious diseases are listed below.

Brucellosis

This disease, which is caused by a bacterium, is also known as contagious abortion. Brucellosis can result in temporary or permanent sterility in females. Abortion is the most common symptom and can occur at any stage of gestation, depending upon the time of exposure to infection with the bacterium. In boars, testicles may become inflamed and permanent sterility may result. The disease is transmitted at mating or by contaminated feed or water. There is no treatment and infected animals should be culled, particularly because brucellosis is transmissible to humans, and the risks of transmission are relatively high under some traditional systems of pig management. Brucellosis appears to be widespread in pig herds in South East Asia and the Pacific Islands.

Coccidiosis

This is caused by organisms known as coccidia, of which there are 13 known infective species in swine throughout the world. They cause damage to the intestinal wall, and are believed to be an increasing cause of diarrhoea in piglets, particularly in confined housing. Affected piglets have grey-green diarrhoea, lose weight and rapidly become dehydrated. Coccidiosis is spread by contaminated faeces and thus good management and regular cleaning of buildings will prevent the disease. Drugs known as coccidiostats are available for prophylaxis and treatment.

Cystitis/nephritis

Cystitis is due to bacterial infection of the bladder and can lead to kidney infection. It is particularly prevalent in sows in confined housing. The disease can be recognised by the presence of pus, and later blood, in the urine. Generally species of *Streptococcus* and *Corynebacterium* are involved and, although large doses of antibiotics can sometimes be beneficial, the disease is usually fatal due to the difficulty of getting the antibiotic to the site of infection. Antibiotic treatment of the penile sheaths of boars and increasing exercise levels for sows by avoiding confinement can help prevent the condition.

Enteric colibacillosis

Diarrhoea caused by the bacterium *Escherichia coli* is the most common cause of death in baby pigs. *E. coli* are normal inhabitants of the intestinal tract, particularly the large intestine. If the pig is subject to stress conditions, they can multiply rapidly in the small intestine, producing toxins and stimulating massive fluid loss into the small intestine, which leads to scouring and dehydration. Scouring due to *E. coli* is primarily a problem in piglets up to 10 days of age. The sow remains healthy and the piglets continue to suckle, but they develop a severe watery diarrhoea and may die. The infection spreads slowly among newborn piglets. Good management is the main preventative measure, i.e. not allowing stressful conditions to develop. Oral antibiotics can be effective if given immediately symptoms are detected. Vaccines are also available which, when given to sows and gilts, help ensure that newborn piglets receive additional antibodies via the colostrum.

Gut oedema is another condition caused by *E. coli*. It affects pigs after weaning, normally at between 8 and 25 kg live weight. The condition is most prevalent in the fastest-growing pigs and the typical symptom is oedema in the region of the stomach and intestines. Pigs begin to stagger and will often die after exhibiting convulsions. The proportion of animals affected varies considerably. It is generally believed that stress is the major cause of gut oedema, particularly overeating of high-protein diets. The disease can be contained by reducing feed intake, increasing fibre levels and lowering the protein content of the ration.

Newer techniques that have proven effective aim to prevent the rapid multiplication of *E. coli* in the first place. These include acidification of the gut, which stimulates the development of favourable bacterial populations (see Chapter 5), oligosaccharides, which bind to *E. coli* and prevent them from attaching to the gut wall, and higher levels of certain trace minerals (e.g. zinc and copper, especially in the chelated form), which stimulate immune response and improve the integrity of the gut lining.

Enzootic pneumonia

This condition is caused by mycoplasmas and is widespread in parts of the tropics. Mycoplasmas cannot live for long outside the body and close contact of infected and healthy pigs is the most common form of transmission. The clinical signs are a raised body temperature and an increased respiration rate accompanied by bouts of coughing. Enzootic pneumonia is a complex condition in which various environmental conditions and stress are involved. The mycoplasmas are resistant to antibiotics and the best policy is to try to keep a herd free from the disease by keeping a closed herd and using boars only from herds that are free of the disease. Antibiotic treatment is useful to prevent the occurrence of secondary infections.

Erysipelas

The causative agent of erysipelas is a bacterium that normally lives in the soil. There are three distinct stages of the disease: acute, sub-acute and chronic. In the acute form sudden death is common. Sick pigs show marked constipation, a high temperature (41–42°C) and a reddish/purplish discoloration of the ears, abdomen and legs. The discoloration often shows up in somewhat rectangular raised patches of skin. The subacute stage is less serious than the acute. Most serious is the chronic stage, which leads to a chronic arthritis, swollen joints and stiffness, and heart damage. Affected joints will be condemned at slaughter.

Once the disease is diagnosed, it can be cured by administering antibiotics as soon as possible. Excellent vaccines are available and a routine vaccination programme is recommended to prevent infection. The disease is transmitted by animal contact or is picked up from the soil. Because erysipelas, like brucellosis, can be transmitted to humans, care must be taken by people who are in contact with infected animals. This means they should wash and disinfect their hands, boots, etc. after contact with infected pigs.

Greasy pig disease (exudative epidermitis)

The disease is caused by infection of the skin with a bacterium, *Staphylococcus hyiens*, and affects pigs aged one to seven weeks of age. Scales form on the skin, which later become crusty, and the skin becomes greasy with matted hair. There is no fever and pigs do not scratch. In the early stages, the condition can be confused with zinc deficiency (see Chapter 5). There are effective medications for dipping infected pigs, and these may be combined with broad-spectrum antibiotic injections.

Mastitis–metritis–agalactia (MMA syndrome)

MMA is common and is the result of bacterial infection in the newly farrowed sow. The three conditions may occur singly or in any combination. Mastitis causes the mammary glands to become swollen and inflamed, and the sow may also have an infected uterus (metritis). In either case there is invariably a lack of milk flow (agalactia) and piglets will become thin and begin to die on the second or third day due to starvation.

A number of stress factors (e.g. inadequate diet, high ambient temperature, nervousness, other disease and retained placenta) may be instrumental in predisposing the sow to MMA syndrome. Everything possible must therefore be done to avoid stress in the farrowing quarters. For sows with MMA, the first essential is to get the milk flowing again and this can be achieved by injections of the hormone oxytocin. Antibiotics can be used to counteract the metritis and mastitis. Adequate exercise and a plentiful water supply during pregnancy are also important in helping to prevent MMA.

Hog cholera

Also known as classical swine fever (in contrast to African swine fever), hog cholera is a highly contagious acute viral disease, and is frequently fatal. The disease is spread by animal contact, contaminated urine and faeces or other body secretions. Initial symptoms are a loss of appetite and a high temperature. The pig may then develop inflammation of the eyes, with a discharge that causes the eyelids to stick together. The pig will also show severe diarrhoea, trembling and lack of coordination, and death will often ensue after four to seven days. There is no effective treatment. If there is a danger of hog cholera infection, a vaccination programme should be instituted.

Porcine babesiosis ('redwater')

This disease is transmitted by the *Rhipicephalus* species of ticks, and particularly affects adult pigs. The prominent features are a high temperature (41–43°C), loss of appetite, weakness, constipation and death within 24–30 hours. There are effective drugs against babesiosis, but the disease will not occur if access to ticks is prevented.

Salmonellosis

Salmonellosis is another enteric disease, caused by the *Salmonella spp.* of bacteria. Pigs are generally affected at around two months of age; they become gaunt, have a high temperature and a foul smelling diarrhoea. There are usually some deaths in a group of infected pigs. An outbreak is often triggered by a stress condition, particularly heavy worm infestation. The disease can therefore be prevented by good management and sanitation. Antibiotics and sulphur drugs will help control an outbreak.

SMEDI

SMEDI stands for stillbirth (S), mummification (M), embryo death (ED) and infertility (I). It is caused by viruses, mainly porcine parvovirus and the enteroviruses. The symptoms vary according to when the sow or gilt becomes infected. If infection occurs during the oestrous cycle and at service, the sow will show a regular or irregular return to oestrus, or if only some embryos die she will produce a very small litter. If infection occurs after 35 days of pregnancy, the foetuses die and dry up and are presented at farrowing as 'mummified' foetuses. The condition can cause a serious decrease in sow productivity within a herd.

There is no treatment, but effective preventative vaccination programmes are available. If vaccines cannot be obtained, all gilts and new animals entering the herd should be given access to farrowing house waste 30 days before breeding. This exposes them to the viruses and stimulates immunity.

Swine dysentery

This disease is caused by a large spirochaete (a spiral-shaped bacterium), and is manifested by severe diarrhoea with reddish-black faeces. Infected pigs rapidly lose weight. The disease is spread by infected dung and can largely be controlled by good hygiene. There are effective antibiotic medications on the market.

Swine influenza

Swine influenza is a highly contagious respiratory disease caused by an influenza virus. It is normally triggered by a stress factor, particularly rapid changes in temperature. Although mortality is low, the disease has important economic consequences due to stunting and reduced live weight gains. The first sign of the disease is normally a cough, with a high temperature and loss of appetite. The disease spreads rapidly, breathing becomes jerky and the hair coat develops a rough appearance. Secondary infection with bacteria may complicate the condition. There is no treatment or preventative vaccine available, but infection can be prevented by good management including avoidance of stress.

Swinepox

Swinepox is a viral disease transmitted either by direct contact or by ectoparasites such as lice. Small red areas (about 1.25 cm in diameter) appear on the skin around the head, ears and ventral surface, which eventually form scabs. There is no treatment for swinepox, but although unsightly it rarely causes serious loss and clears up after a short time.

Transmissible gastroenteritis (TGE)

TGE is a viral disease causing acute diarrhoea, vomiting and early death in young piglets. It also affects older pigs, causing diarrhoea and vomiting, but rarely death. There is no treatment. Infected pigs can be isolated, or killed and buried. After infection, the whole herd is likely to be immune.

Non-specific diseases

Abscesses

Abscesses can occur as the result of any irritation, inflammation or wound that allows bacteria to get into the body, normally strains of *Staphylococcus* or *Streptococcus*. The body of the pig reacts to the invasion of bacteria, and a pocket of pus is walled off from the body. Abscesses are seen as swellings or lumps, are often hot to touch and, in time, they develop a soft area, which can be lanced and drained. They may be superficial, or they may form deep within the body, where they can cause lameness, interfere with breathing or swallowing, or may not be discovered until slaughter.

As abscesses are painful and can markedly depress performance and reduce carcass value, every effort should be made to minimise the possible causes in a piggery. Preventative measures include removing any sharp or rough objects from pig pens, ensuring the floors are not too rough, especially for piglets, making sure that injection equipment is sterilised and providing good overall sanitation. Abscesses can be treated with antibiotics, but this is not always effective.

Gastric ulcers

Ulcers tend to occur as a response to stress in pigs of all ages, and are particularly prevalent in genetic strains bred for fast growth with a thin covering of back fat. The nature of the ration can also be a factor, with ulcers occurring more often in pigs fed on finely ground, high-energy concentrate diets. There may be no specific external symptoms, unless haemorrhaging occurs. Otherwise pigs show a lack of appetite, will huddle together and become thin. Mortality varies according to the extent of the ulceration. There is no specific treatment apart from reducing stress. Changes in the ration involving an increase in fibre levels is often useful in ameliorating the condition.

Intestinal haemorrhage syndrome ('red-gut')

This condition occurs in pigs of 40 kg or over, and the first sign is usually death. A post-mortem will show the presence of blood in the intestinal tract, without any associated inflammation of the intestinal wall. The cause of this condition has yet to be established, but it is believed that environmental factors may play a part.

New viral diseases

During the past decade, several new viral diseases have appeared in the pig herds of Europe and the USA. These include porcine reproduction and respiratory syndrome (PRRS), also known as 'blue ear' disease, porcine dermatitis nephropathy syndrome (PDNS) and post-weaning multisystemic wasting syndrome (PMWS). Great care needs to be exercised to prevent these diseases spreading into industries in the developing world by minimising imports and by screening to ensure animals that are imported are free from these diseases.

Mycotoxins

Mycotoxins are a diverse group of chemicals that are harmful to pigs. They are secreted by certain fungi that grow on crops in the field, during transport and in storage. Tropical environments are often particularly favourable for the growth of these fungi. There are four main groups of mycotoxins:

- *Aflatoxins*: These are produced in species of the *Aspergillus* fungus and are highly toxic. They increase susceptibility to disease (depress immune function) and may affect liver function. The first signs of aflatoxin ingestion are reduced appetite and slowed growth rate.
- *Deoxynivalenol (DON or vomitoxin):* Produced by *Fusarium graminareum* – more than one part per million in pig feed will cause reduced feed intake followed by vomiting.
- *T2:* A product of *Fusarium sporotrichoides,* this causes infertility with lesions in the uterus and ovaries.
- *Fumonisin:* This is produced by a range of *Fusarium* species. Lung oedema and liver necrosis are the most common symptoms in pigs.
- *Zearalenone:* Another product of *Fusarium graminareum.* Best known for its role in oestrogenic syndrome in pigs. Females exhibit reduced feed intake, swollen vulvas, an enlarged uterus and shrunken ovaries. Young males become 'feminised', with enlarged nipples and shrunken testicles.

Tests for mycotoxins in feedstuffs are expensive and not always available in developing regions. Any ingredients suspected of contamination should be immediately withdrawn from the feed. Mycotoxin binders (see Chapter 5) can also be effective in detoxifying the feed.

Diagnosis chart

Symptoms	Possible disease or condition
Abortion	Brucellosis. Redwater. Any disease causing a high body temperature.
Agalactia	MMA. Ergot in feed.
Anaemia	Roundworms. Redwater. Iron deficiency. Trypanosomiasis.
Appetite (loss of)	African swine fever. Hog cholera. Redwater. Ulcers. Swine influenza.
Breathing (distressed/rapid)	African swine fever. Swine influenza.
Constipation	Redwater.
Coughing	Roundworms. Lungworms. Enzootic pneumonia. Swine influenza.
Dehydration	Coccidiosis. Enteric colibacillosis.
Diarrhoea (scours)	Roundworms. Coccidiosis. Enteric colibacillosis. Hog cholera. Salmonellosis. Swine dysentery. Transmissible gastro-enteritis.
Discharges	
anus	Anthrax. Intestinal haemorrhage syndrome.
eyes	Hog cholera.
vulva	MMA.
Emaciation	Roundworms. Ulcers. Coccidiosis. Tapeworms. Salmonellosis.
Infertility	Brucellosis. SMEDI.
Nervous signs	African swine fever. Hog cholera. Rabies.
Salivation	Foot and mouth disease. Rabies.
Skin conditions	Mange. Lice. Fleas. Erysipelas. Swine pox. Greasy pig disease. African swine fever.
Sudden death	Anthrax. Acute trypanosomiasis. Acute erysipelas. Intestinal haemorrhage syndrome.
Swellings	
face	Foot and mouth disease. Abscesses.
body	Abscesses. Anthrax. Gut oedema.
feet	Foot and mouth disease.
udder	MMA.
Temperature elevated	All bacterial, protozoal and viral diseases.
Urine (discoloured)	Cystitis. Redwater.

8 Practical management

General considerations

Practical management of pigs has, without doubt, a marked effect on the profitability – or otherwise – of a pig producing enterprise. Through good management, the producer can integrate all the factors relating to the well being and productivity of his animals in the most cost-effective way. Good management invariably depends on a high degree of attention to detail.

Stockmanship and pig handling

Stockmanship concerns the personal relationship and rapport that develops between a stockman and his pigs. Pigs are very responsive to quiet, calm conditions. The good stockman is firm, but understands the needs of his pigs. He will always be on the watch for any irregularities in the behaviour or condition of a pig and thus learns to anticipate problems before they arise.

Pig handling is an adjunct of good stockmanship, and should be carried out in as calm a way as possible. It is particularly important to upset pigs as little as possible when they are being caught or driven. When catching a suckling pig, it should be approached from behind and lifted suddenly by holding one or both hind legs above the knee joint or hock. It can then be held around its shoulder. Bigger pigs can be caught by grasping the pig behind its shoulder with both hands, lifting it off the ground and then holding it firmly against the body. Pigs cannot be led by the head but must be driven from behind. The natural instinct of the pig is to head for any gap and the skill is therefore to make maximum use of any walls, solid hurdles or pig boards to ensure that the only gap is in the direction you wish the pig(s) to proceed.

Occasionally it is necessary to restrain a sow or boar for treatment purposes. The best method is to make a slipknot in a 2 m length of rope, which is used as a noose around the pig's mouth, encircling the snout close up against the juncture of the upper and lower jaws. When in place draw the noose tight. If necessary, twist the surplus rope around a railing or pole, but normally a man holding it firmly at about 35° to the horizontal will be able to hold the pig, which will always pull against the pressure and stand firm for any treatment. For quick release, do not pull the rope right through when completing the knot, but leave it in a loop. When the loose end is pulled, the knot will fall apart.

Good hygiene

The importance of good hygiene in a piggery cannot be over-emphasised. Not only does it help to reduce the incidence of disease in the pigs, but it also has a beneficial effect on staff morale by improving the working environment. A cleaning programme should allow for the periodic emptying and resting of each house. For the farrowing house, this should consist of a week's rest in between each batch of farrowing sows. For fattening pens, a five-day break after each group of fatteners is adequate. Immediately after it is empty, each pen should be thoroughly scrubbed and cleaned, washed and soaked in disinfectant and allowed to remain dry for the remainder of the period that it is empty. This 'all in, all out' system is very important in preventing the build up of disease and infection.

Boars

Young boars

Although it is may be said that 'the boar is half the herd', management of boars is frequently one of the most neglected aspects of a pig enterprise. Young boars should be exercised daily so they can get used to their stockman and the sights and smells of the piggery. They should be fed to provide for continued growth, but should not be allowed to become fat and sluggish. A ration of 2–2.5 kg of a standard sow and weaner meal each day is usually adequate.

A boar should be ready for service when he is 7.5–8 months of age. In the first instance, he should be introduced to a small sow that is well on heat and standing firm. This will prevent him becoming frustrated and allow for a successful first service and a favourable sexual experience. At no time should a young boar be introduced to a group of females, as the bullying that may ensue can adversely affect the boar's libido for some time afterwards. For the first few months of his breeding life, the boar should not be overworked and a maximum of two services per week should be the rule.

Mature boars

Maintaining a mature boar in a lean, hard condition that is ideal for service requires good stockmanship. A balanced ration should be fed at a level to prevent the boar becoming either too fat or too thin.

Research has shown that isolating mature boars (i.e. boars of more than 10 months) from female pigs can severely reduce their level of sexual behaviour. Thus, boars should be housed within sight and sound of sows and gilts. Under tropical conditions, it is imperative that boars are protected as much as possible from high temperatures, otherwise they will suffer from depressed libido and sperm production. Their accommodation therefore must be designed with this in mind (see Chapter 6). Spraying them with water, by hand or using sprinklers timed to operate intermittently, and allowing access to wallows can all be beneficial in reducing heat stress and sunburn.

Mating management

The need for and design of service pens is explained in Chapter 6 and Fig 40. An Australian study (Hemsworth, 1978) has confirmed the benefits of a service pen, finding that boars mating in their own pens displayed lower levels of sexual behaviour than boars mating in the service pen. This resulted in a lower percentage of gilts being mated in the boar pen. Problems can also arise when large boars are too heavy for the sow when they can cause her to collapse at service. This can be prevented by restraining the sow in a service crate (Fig 55), where wooden supports take the weight of the boar's front legs. Service performance can also be maximised by conducting mating early in the day or in the evening, when temperatures are generally lower.

To avoid overworking boars the ratio of sows to boars should not be too large. The standard recommendation is one boar to 20 sows and gilts. However, this must be related to peaks of service activity. In units where boar numbers are low, it may be beneficial to allocate two days each week instead of one for weaning sows. The sows will then come back on heat at different times and the boars will have time to recover between services and will continue to produce fertile semen.

There is a great variation between boars in their ability to work consistently, but in general terms overworking a boar will reduce the quality and quantity of sperm produced, leading to small litters and more sows having to return to service because they are not pregnant. By ten months of age, a well-managed boar should be able to cope with five or six services per week and still maintain good fertility levels. On the other hand, a boar should not be under-worked, and intervals of more than 14 days between services can lead to reduced litter sizes because the boar becomes less fertile.

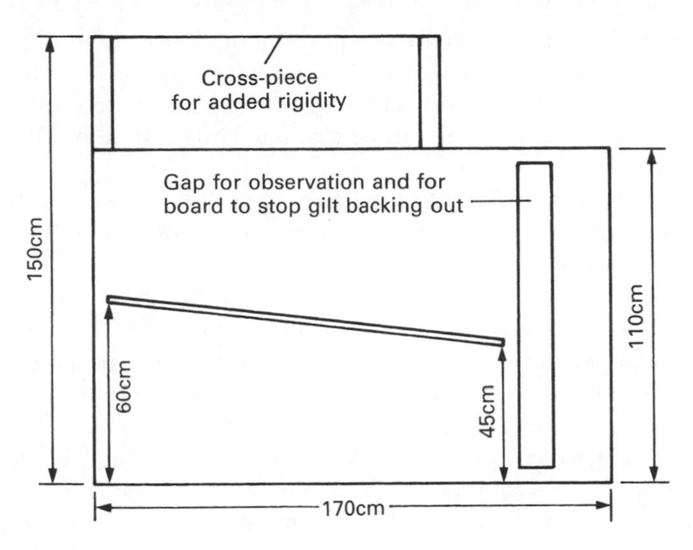

Fig 55 *Dimensions for a service crate (side view)*

Lameness

Surveys from many countries have shown that some 20 per cent of boars are culled for foot problems and lameness. Prevention is better than cure, and problems in the feet can often be avoided by ensuring floors are kept clean and are not too rough. New concrete is particularly damaging to skin and feet because of its extreme alkalinity. Supplementary biotin in feed (see Chapter 5) and the use of footbaths containing 5 per cent formalin or 5 per cent copper sulphate solution can help to harden the claw horn.

Sows and piglets

The young sow (gilt)

Gilts selected for breeding from within the herd should be moved out of fattening pens at five months of age into groups in well-bedded pens or yards, or even outside paddocks. They should be fed to continue growing without becoming over fat. A ration of 2–2.5 kg of a standard sow and weaner meal a day is usually adequate.

When gilts are brought into the herd from elsewhere, they need special care and attention and a period of 'integration'. Ideally, they should be purchased in batches at least six weeks before they are due for service. They should be examined carefully to see that they are not damaged and should be encouraged to eat normally in their new environment. The gilts can be housed in pens or yards, and should be given access to dung and waste from the farrowing house, as this helps them to build up immunity to diseases. Thereafter, they can be gradually integrated into the main gilt/sow herd.

The main objective with gilts is to encourage them to reach puberty as soon as possible because: a) the stockowner can dispose of any that are not showing breeding activity at an early stage, b) the stockowner has access to a choice of young, sexually active gilts, and c) gilts will be at second heat or more at their first mating, thereby increasing the size of their first litter.

Onset of puberty can be stimulated by introducing a boar to the gilts for 10 to 15 minutes every day. The boar should be at least 9 or 10 months old and actively working so that he is producing the necessary odours (pheromones) to stimulate the gilts. Gilts are most sensitive to boar stimulation at 145–170 days of age. If gilts show reluctance to come on heat for the first time, they can be encouraged by: a) mixing them with strange gilts, b) moving them into new accommodation, and c) taking them for a ride in a truck or trailer.

Gilts of exotic breeds should be over 120 kg bodyweight when served and in good active condition. *Ad lib* feeding for 14–21 days before service helps stimulate the number of eggs and consequently increases litter size. The boar used on the gilts should not be too heavy, or he may injure them. Alternatively, a service crate (Fig 55) can be used. However, successful service using a service crate is not always easy to achieve, and it is generally best to try and avoid too much disparity in weight between the boar and the gilt.

Lifetime productivity

Although much emphasis is placed on output of the sow in terms of numbers of piglets reared per litter, the ultimate measure of her productivity and profitability is the number of healthy weaners reared in her lifetime. Management strategy must therefore be aimed at maximising this output within the constraints of each enterprise. However, before considering the separate phases in the breeding cycle of a sow, namely mating and pregnancy, farrowing, lactation and weaning, it is essential to realise that they cannot be dealt with in isolation. Each phase will impact on the other and, with feeding for example, there can be large carry-over effects from one phase to the next.

Opinions on the best way to feed sows have differed over the years, but it is now generally agreed that the best guideline is to minimise bodyweight fluctuations over the entire reproductive cycle. This means providing relatively low amounts of feed during pregnancy and high (*ad lib*) levels during lactation. This not only prevents the sow becoming too fat and heavy at farrowing, thereby minimising farrowing problems, but also prevents her becoming too thin at weaning, thereby avoiding delays in oestrus and rebreeding. At the same time, allowance should be made for continued growth and weight gain for at least the first three to four reproductive cycles. In this respect, it is often recommended that a sow should gain some 12 to 15 kg of bodyweight from weaning to weaning for the first four parities (see Fig 56).

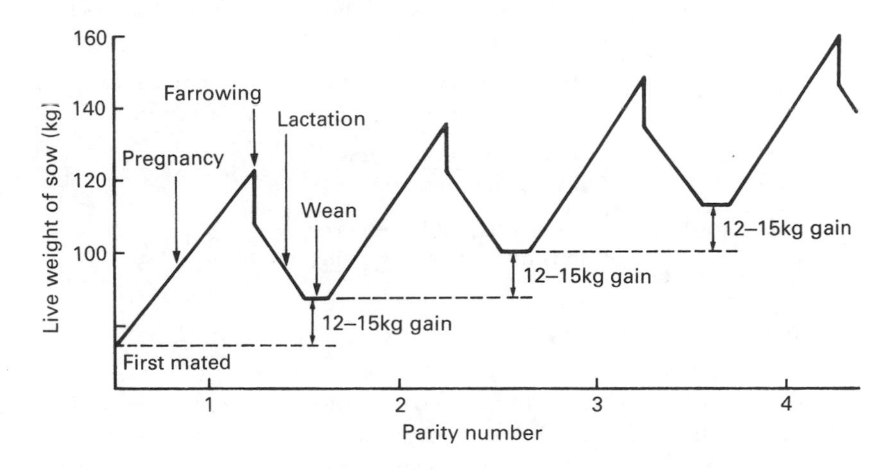

Fig 56 *Recommended bodyweight increases in sows over successive parities*

Work by Professor Whittemore and his colleagues at Edinburgh University (Whittemore, 1987) has indicated that, particularly for modern, leaner sows, the state of the sow's body fat reserves is likely to be more important for reproduction than weight gain. A 160 kg sow, for instance, may be a small-framed animal carrying a large amount of fat, or a large-framed animal in very poor condition. As a consequence, a condition scoring technique has been developed for the sow and is widely used as a means of monitoring the body reserves of sows at strategic points throughout the breeding cycle. Sows are scored on the basis of a combination of visual appraisal and physical feel at various points such as the hip bones, tail setting, loin area, backbone and ribs. This provides a useful guide to the management and feed requirements needed to achieve certain condition scores at set points throughout the breeding cycle. A simplified description of the sow condition scoring system is given in Table 12.

A change in approach from feeding according to set requirements to a more flexible system of feeding according to the body condition of the sow is particularly appropriate in the tropics, because it allows for more

Table 12 Condition scoring system for the sow

Score	Description	Assessment of fat cover over backbone, transverse spinal processes and hips	Appropriate score for stage of the breeding cycle
1	Poor	Visually thin, hips and backbone prominent, no fat cover	
2	Moderate	Bones felt easily without pressure on palms	Sow at weaning
3	Good	Bones only felt with firm palm pressure	Sow prior to farrowing
4	Fat	Bones cannot be felt with firm palm pressure	
5	Grossly fat	Bones cannot be felt even by pushing with a single finger	

Source: DEFRA (2004)

effective use of less conventional feeds of lower nutrient density. The adequacy of these feedstuffs can be assessed by the responses in sow condition, particularly when they are fed to dry sows.

Mating management

The crucial points in effective mating management are to recognise the onset of oestrus and then to serve at the correct time. Ensuring sows are served during the period of peak fertility (see Fig 32) will reduce the likelihood of sows returning to service or producing small litters. Characteristic signs of oestrus under practical conditions are:

- swelling and reddening of the vulva (not always obvious);
- confined sows will spend less time lying down;
- sows will be more alert and the ears will stand up in prick-eared breeds;
- sows will allow themselves to be mounted by the boar or other sows;
- sows will give a characteristic 'grunt';
- sows will exhibit the 'standing reflex', i.e. they will adopt an immobile posture when pressure is applied to the back.

Ideally, sows should be served in a specially designed service area (see Chapter 6) and each service should be quietly supervised by a stockman to ensure that mating has been effective. In the tropics, sows can quite easily be served outside under a tree and this is generally preferable to serving in the boar pen if a special service area is not available. Sows should be served during the coolest time of day.

Under careful management, a single mating is perfectly adequate for good conception rates and large litters. However, it is generally safer to serve sows twice, once approximately 12 hours after the sow is first seen on standing heat, and the second time a further 12 hours thereafter. In some units a triple mating procedure at 12-hour intervals is used. This can be considered if unsatisfactory results are obtained with a double mating system.

The pregnancy period

The sow's ration should be reduced the day after service, because high feeding levels during early pregnancy arc associated with loss of embryos. Embryo loss can also be minimised by preventing any undue stress and this includes avoiding high environmental temperatures.

Sows should be monitored during early pregnancy and any that return to oestrus can be served again. If available, a pregnancy detector can be used to confirm pregnancy. There are a variety of instruments available, most of which use ultrasound to detect increased blood flow to the uterus and foetal heartbeats. Failing to check for empty sows can be costly, because by the time they are finally discovered they should have been due to farrow.

The pregnancy period is an ideal opportunity to feed bulkier feedstuffs. The sow's appetite will be well above her requirement for concentrated nutrients, and the 'fill' effect of bulky feeds keeps her more content. As long as the ration is balanced and sow condition is maintained, considerable savings in feed costs can be achieved. Once a day feeding of dry sows will also save cost.

Farrowing

Farrowing, or parturition, is undoubtedly one of the most critical stages in the whole reproductive cycle. Problems at this stage very quickly lead to high mortality rates and reduced efficiency in both sows and piglets. The skill of the stockman is to recognise what constitutes a normal farrowing so that any departures from normal can be detected and corrected.

Two or three weeks prior to farrowing, the sow should be wormed to minimise the possibility of passing worms to the newborn piglets. One week before farrowing, she should be washed with soapy water to remove dung, then moved to her farrowing quarters. She can also be sprayed with insecticide at this stage to remove any mites and lice.

Constipation must be avoided at all costs as this interferes with the farrowing process. If the herd has a tendency to constipation, the diet should be supplemented with bran, green leaves or extra fat during the last week of pregnancy.

There are two sure signs that farrowing is about to commence: a) the sow will show increased restlessness (in direct contrast with the peaceful animal of late pregnancy) and she will start to make a nest by rearranging her bedding, and b) milk can be withdrawn from the udder about 12 hours prior to the start of farrowing.

Although not essential, it is recommended that all sows farrowing under intensive, confined conditions should be supervised by an experienced stockman. Clearly, this is not practical for extensive conditions, but such sows are generally better able to farrow unaided.

Two major problems may be encountered during farrowing. Firstly, the process may become painful, particularly for gilts if the piglets are relatively large. This can result in the dam savaging her litter. Should this be observed, all piglets should be removed until farrowing is complete and then the piglets can be reintroduced slowly under supervision. Exercising the sow can help, but it may be necessary to muzzle (e.g. using part of the leg of an old gumboot) or tranquilise the sow.

The second problem is delayed farrowing. After farrowing has started, any time delay of greater than 30 minutes between presentation of piglets probably means that a piglet has become stuck in the birth canal. In this case, the stockman should disinfect and soap his arm and then gently work it into the birth passage to release the piglet. The sow will often stop straining in this situation, but an injection of oxytocin will usually get contractions going again and farrowing will then proceed normally. The only certain sign that farrowing has been completed is the passage of the afterbirth.

Stillbirths

Surveys in Europe have shown stillbirth rates ranging from 1 per cent to more than 12 per cent of all piglets born, so, in some cases, it is the single most important cause of piglet mortality and seriously depresses productivity. It is believed that over 70 per cent of stillbirths are caused by asphyxiation or weakening during the birth process and that some 70 per cent of these occur in the last 30 per cent of piglets born. This is due to the design and length of the sow's uterus (see Chapter 2), which means that piglets from the far end have to travel some 30 cm after their umbilical cord has ruptured. The intact cord supplies the piglet with oxygen from its mother's blood, but once the cord is broken, the piglet will start gasping for oxygen. It then has five minutes to get to the outside before it will stop breathing.

Many factors can cause delays in the farrowing process and thereby increase the incidence of stillborn piglets:

- *Large litters:* Because the whole process will take longer, the last piglets to be born have a higher risk of being suffocated.

- *Very small litters:* These are liable to provide insufficient stimulus to the sow to start effective uterine contractions.
- *Older sows:* After five or six litters, uterine tone declines, resulting in a less efficient birth process.
- *Genetic make-up of the sow:* Variation can occur between families in the efficiency of farrowing. If a strain of pigs is identified that regularly has a high incidence of stillbirths, then they should be culled.
- *Nutrition*: Sows should not be over fat or too thin with insufficient reserves of energy.
- *High ambient temperatures:* Sows will tend to tire more quickly at high environmental temperatures.
- *Piglet diseases:* Dead or mummified piglets can slow down the birth process and increase the risk of suffocation for the live ones.

If stillbirths become a problem in the herd, every attempt should be made to minimise the effect of any or all of the above factors.

The newborn piglet

The main factor affecting the survival of newborn piglets is that they are born with very limited supplies of energy. In addition, the stress of the birth process can use up their small energy reserves. Management of newborn piglets should therefore seek to minimise additional stress and ensure they have immediate access to the sow's milk.

The first 72 hours after birth are critical for the baby pig. During this period the colostrum of the sow has a high content of antibodies and the piglet intestine is able to absorb intact proteins. As the piglet has very little of its own resistance to disease, it is essential that it gets a good suck of colostrum and acquires passive immunity from the sow. Failure to take in sufficient colostrum will invariably result in the pig succumbing to infection before it can develop active immunity of its own. Once the piglet has established a teat position, which normally occurs in the first 24 hours after farrowing, it will retain this position for the remainder of the suckling period. A healthy, good size piglet will get to its feet minutes after birth and instinctively try to reach the sow's udder. On average, such piglets obtain their first suckle of colostrum within 45 minutes of birth. The weaker and lighter piglets (less than 1.0 kg at birth for exotic breeds) are less able to reach the udder, are generally less competitive if they do reach it, and are the most liable to die if they are not assisted. As long as milk production is adequate, a sow will suckle her litter every 60 to 90 minutes.

Management techniques to maximise piglet survival include the following:

- *Maintaining temperatures:* Ensuring the piglet is not exposed to cold or draughts prevents it having to use energy to keep warm.

- *Cross-fostering:* This is a well proven and widely adopted technique. When sows farrow within a few hours of each other, litters are evened up by transferring piglets from sows with large litters to those with smaller litters, thereby increasing the survival of smaller piglets from larger litters. When cross-fostering, it is advisable to transfer the bigger piglets on to the foster mother. Care must also be taken that when piglets are transferred on to different sows, the teats are not too big for them to be able to suckle successfully.
- *Split suckling:* After the larger piglets in the litter have had a good suckle of colostrum (approximately 1 hour at the udder), they are shut away in a warm box for the next two hours. This gives the remaining smaller piglets access to the udder to get their share of colostrum. This can be repeated twicc in the first 18 hours.
- *Supplementation:* Small and weak piglets can be given additional colostrum that has been stored. Colostrum can be obtained by milking sows before they farrow or during farrowing, and can be stored in a freezer. It can then be given to needy piglets via a bottle or, more effectively, via a stomach tube. This consists of a soft tube of around 25 cm in length, which is inserted down the throat directly into the stomach.
- *Extra fat in the sow's diet:* The addition of some supplementary fat in the diet of sows for seven to 10 days before farrowing can increase the fat and energy content of the sow's milk and may slightly increase piglet energy reserves and vitality.

Other routine operations that should be carried out at this stage:

- Navel cords should be immersed in a dilute solution of iodine within 24 hours of birth. Any long navel cords should be trimmed to about 5 cm long. Iodine treatment prevents infection entering through the navel that could cause joint-ill (swollen joints). Piglets with joint-ill fail to thrive and may die.
- Eye teeth should be clipped within 24 hours to remove the points. This process helps to prevent teat and udder injuries.
- Within their first three days of life, housed piglets should receive supplementary iron to prevent anaemia. If possible, iron should be given by injection. Otherwise, soil enriched with ferrous sulphate should be placed in the pen.
- Ear notching (or another system of identification) can also be carried out at an early stage.

The lactating sow

Contented sows make the best mothers, and management should be aimed at keeping sows as comfortable as possible. Directly after farrowing, the sow should be examined for any disease problems, particularly

MMA (see Chapter 7) and, if symptoms are seen, she should be treated immediately. She will need access to plenty of fresh water as she can lose up to eight litres of body fluids from the uterus and associated tissues over a four-to-six hour period at farrowing. Failing to provide sufficient water at this stage can lead to agalactia.

If the sow is not over fat at farrowing, she should be fed at a rate that will minimise weight loss during lactation. Unless she has a very small litter, this generally means feeding as much as she will eat, twice daily. However, it is important to increase feed gradually over the seven days after farrowing, as overfeeding at this stage can put the sow off her feed. If a sow continues to show significant weight loss in spite of feeding to appetite, it may be necessary to feed her more frequently. Giving wet rather than dry feed, particularly in hot climates, also helps to stimulate appetite. Lactation puts a heavy demand on the sow, so care must be taken to ensure that environmental temperatures remain within her thermoneutral zone (see Chapter 2).

In tropical regions, it is recommended that the lactation period is no shorter than five weeks and, in certain cases, it may be cost-effective to extend it further. Although in developed countries, four- and three-week weanings are often practised, the lack of quality creep feed for the young pig in developing countries means that attempts to wean earlier than five weeks have generally been unsuccessful.

Creep feeding

The quantity of the sow's milk reaches a peak at about three weeks after farrowing. Thereafter the piglets must be provided with some solid feed. 'Creep' feeding is so called because the feed is traditionally given in the piglets' nest or creep where it is not accessible to the sow. Creep feeding has several benefits:

- it helps the piglet's digestive system to adjust to the change from milk to a solid diet, thereby minimising digestive upsets and checks in growth rate at weaning;
- it helps reduce the drain of nutrients from the sow, minimises her bodyweight loss in later lactation and leaves her better prepared for breeding again;
- it attracts the piglets away from the vicinity of the sow and reduces the chances of being crushed;
- piglets with access to creep feed grow faster than those having sow's milk alone.

Creep feed should be highly digestible and palatable and formulated to meet the nutritional needs of young pigs prior to weaning and up to eight weeks of age. Ideally, it should be based on skimmed milk powder with added unsaturated fats and good quality non-milk proteins. However, milk products tend to be scarce and expensive in the tropics (and are usually reserved for humans) and creep feeds tend to be of relatively low quality. Simple diets based on maize and soyabean meal can be very effective. Creep feed should be offered to the piglets first at about seven days of age in very small amounts so they can get used to eating solid feed. Quantities can then be increased according to appetite up to weaning. The feed should always be offered fresh on a 'little and often' basis and a separate supply of fresh water for the piglets will help stimulate their feed intake.

Weaning the sow

At weaning, the aim is to dry off the sow, stimulate her to exhibit oestrus and get her to conceive again as soon as possible. At the same time, she should be induced to release a large number of eggs at ovulation as the first step towards a large litter at the next farrowing. Under normal conditions, removal of the sow from her litter and the cessation of suckling will trigger ovarian activity and oestrus will occur within four to seven days. There are several management strategies that will help ensure timely oestrus:

- move the sow into a house where she can hear, smell and have contact with a mature boar;
- house her with or near other newly weaned sows, since other sows coming into oestrus will stimulate the process. Newly weaned sows should not be mixed unless/or until they are used to each other;
- allow the sows to exercise – the cubicle house design (see Chapter 6) is ideal for this;
- feed the sows at a high level after weaning until oestrus occurs – this is particularly beneficial for young sows after their first litter and for thin sows.

Extensive sow systems

The types of sow kept on extensive systems (indigenous, indigenous x exotic, or exotic hybrids) are generally hardier and exhibit better mothering ability than exotic sows bred for confined conditions. Therefore, they tend to need less supervision. Nevertheless, they will respond to good management and stockmanship and the following strategies will help to maximise productivity:

- provide adequate deep shade and wallows;
- site farrowing arks or nests sufficiently far apart to allow sows to establish their territory at farrowing time;
- provide bedding in the farrowing arks;
- gain the confidence of sows so they can be handled easily during critical periods such as farrowing time;

- spread bulky feeds over a large area so all sows can access the feed (for example, if cassava roots are placed in a heap, the dominant sows will guard them and prevent more timid sows from eating);
- provide an adequate boar-to-sow ratio (1:15);
- adopt a rigorous worming programme, unless sows can be moved to new land on a regular basis.

Culling of sows

Maintenance of overall productivity in the herd includes having a culling policy so that sows are removed at the correct time. In the majority of cases the reasons for culling sows will be obvious, e.g. lameness, other injury, farrowing problems, poor litter size, poor mothering ability and low fertility. Even sows that normally produce good litters will start to decline in performance after the fourth litter and this will usually become noticeable by her tenth litter. If she is producing well, a good guide is to allow her to remain in the herd until her performance falls below the average of the gilts in the herd. At the same time it is important to have a supply of pregnant gilts available to replace sows that need to be culled.

The once-bred gilt

This is a relatively new system of producing weaners, and could be useful in the tropics. Gilts are served at an early age and then slaughtered after they have weaned their first litter. The carcass of the gilt is very similar to that of the heavy hog (see 'Growing and finishing pigs', below). The system exploits the phenomenon known as pregnancy anabolism, whereby the deposition of maternal tissue during gestation is very efficient in terms of feed utilisation. Approximately 50 per cent of the total weight gain during pregnancy will consist of body gain, and the other 50 per cent will comprise the products of conception. The 'once-bred gilt' system can be considered as an adjunct to a conventional sow-based system. Some older sows will generally be required in the herd to maintain immunity against reproductive diseases. Nevertheless, it could prove a useful way of producing cheaper meat in a pig enterprise.

Weaners

The three-week period immediately after weaning is a critical one for the young pig, because it is simultaneously confronted with a number of stresses:

- psychological trauma of separation from its dam;
- social stress of mixing with pigs from other litters;
- stress of settling into the strange environment of a new pen;

- major change in diet associated with the loss of sow's milk;
- loss of the supply of immunoglobulins and immune protection provided by the sow's milk.

Practical management at this stage aims to achieve the transitional process of weaning without any growth checks or mortality and to start to accelerate the growth of the pigs. This is important because the young pig's potential for efficient growth at this time is high, and any factor which reduces growth rate can prove costly. However, the task is complicated by the fact that a major change in diet often leads to digestive scours, which in turn provides an ideal environment for multiplication of *E. coli* and problems of gastroenteritis and bowel oedema (see Chapter 7).

Immediately before weaning, pigs should be handled and disturbed as little as possible. Any essential management tasks should be carried out at least two weeks before weaning. Piglets should be encouraged to consume as much creep feed as possible before weaning and should continue to be fed creep feed for at least two weeks after weaning.

Ad lib feeding is recommended to capitalise on the high growth potential of the young pig. However, as soon as any scours are seen, feed intake should be restricted. If the problem persists, consideration should be given to increasing roughage levels and reducing the protein content of the diet. Although the prophylactic use of antibiotics is now restricted, various feed additives can help control the development of bacterial scours (see Chapter 5).

At weaning, batches of pigs should be selected from different litters according to bodyweight. Excessive fighting after mixing can be avoided by supervising the pigs for a few hours or, alternatively, by smearing the pigs with oil. Pigs can be retained in their litter groups, which minimises fighting but results in batches of different size and tends to give rise to uneven growth rates.

Excessive variations in ambient temperature within the house should be avoided (see Chapter 6). Pens should large enough to allow the pigs to spread out and keep cool. Wet floors or wallows should be provided if excessive heat is a problem (Fig 57).

Ideally pigs should be in good condition at weaning, so that they possess some fat reserves to help combat stress. Poor litters should be allowed to remain with the sow for longer than five weeks. A technique of 'split' weaning can also be beneficial, whereby the large piglets are weaned first and the small ones are left with the sow for a further week. Fresh water must be available at all times as it encourages higher intakes of feed.

Growing and finishing pigs

By eight or nine weeks of age, the growing pig has passed the stresses of weaning, and its digestive system will be able to deal with a range of

Fig 57 *Plenty of space and wet floors help pigs to keep cool in hot environments*

protein and energy sources. Some 80 per cent of the feed used in a pig unit is consumed by the growing and finishing pigs, therefore the efficiency of feed utilisation during this phase is a crucial factor affecting profitability.

Clearly, the management system must relate to the specific objectives of each unit and these may range from home consumption of as cheap a carcass as possible, to the production of sophisticated bacon products. These considerations dictate the type of pig to be produced and management must be geared to optimise performance. The small-scale rural producer will want to use mainly cheap, low-quality feedstuffs, but should remember that the growing pig has only a limited ability to digest and utilise fibre. Too much bulky feed may depress growth to such an extent as to render its inclusion uneconomic.

A larger, more commercial operation will appoint different priorities to factors such as feed conversion efficiency and feed cost per pig, growth rate, carcass leanness and grading according to the relative economic advantages they confer in the whole production unit. Traditionally, the main categories of slaughter pigs in commercial industries are: a) porkers, which are slaughtered at up to 65 kg live weight, b) baconers, slaughtered at 70–90 kg live weight, and c) heavy hogs, slaughtered at 90–140 kg live weight. In general terms, the heavier the pig is at slaughter, the cheaper is each kilogram of carcass meat that is produced.

Feeding systems

Several different feeding systems can be adopted and the relative merits of these are discussed below. The importance of water to the growing pig in hot climates cannot be over-emphasised. The art of successful fattening is to stimulate maximum feed intake. The absence of a regular supply of clean, fresh water is often the first factor that limits feed intake. Where possible, periodic testing of water supplies for contaminants (both mineral and microbial) is recommended.

Restricted or ad lib *feeding*

In the past, fattening pigs have generally been fed on restricted regimes to avoid producing carcasses that are excessively fat. However, modern exotic strains can grow through to slaughter at younger ages without getting too fat, and can therefore be fed *ad lib.* Thus the decision on whether it is necessary to restrict feed intake will depend on the circumstances of each unit. The advantages of *ad lib* feeding are that it saves labour, gives faster growth rates and produces a larger carcass weight at a given age. In general, it will result in more contented pigs and avoids problems of competition at the feed trough. Care must be taken that *ad lib* feeders are designed to operate efficiently in order to minimise waste. A suitable, relatively cheap and efficient feeder has been developed in Africa (Fig 58). If it is necessary to limit nutrient intake, dietary fibre levels can be increased. Obviously, various combinations of *ad lib* and restricted feeding periods can be used in order to yield the most desirable result. For instance, feeding *ad lib* up to 50 kg bodyweight and then restricting intake until slaughter is a system in common use.

Feeding in a trough or on the floor

Feeding on the floor makes more economical use of floor space, and reduces bullying and competition. However, although feeds should be restricted to what is consumed in 20 minutes, it leads to more waste, a greater problem with dust and, if scouring occurs, greater danger of transmission of disease than occurs when pigs are fed in the trough. If pigs are fed in troughs, adequate trough space must be provided (see Table 11). Frequency of feeding in troughs can also influence the efficiency of feeding. Research has shown that by feeding twice a day rather than once, small improvements are achieved in growth rate, feed required to slaughter and killing-out percentage (see below). Single-space feed hoppers are a relatively new system that is claimed to improve feed intakes and growth rates. With these, pigs penned in groups have access to the feed hopper one at a time, so are protected from interference or competition from their pen mates.

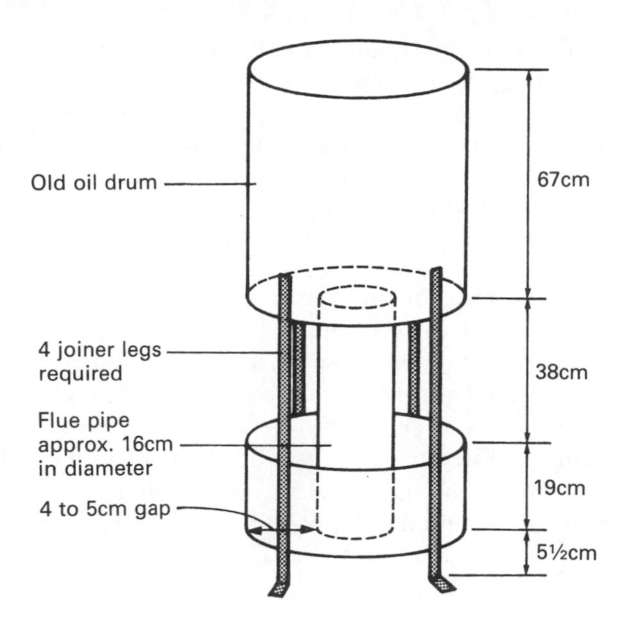

Fig 58 *A self-feed hopper made from an old oil drum*

Wet or dry feeding

There are two main types of wet feeding systems. In the first, the feed is diluted to the extent that it can be delivered to the pens by pipeline or continuous trough (river system). In the second, water is added to feed in the trough. The first system is only appropriate in larger units, where the mixing and delivery can be semi-automated. Trials in Zimbabwe have shown that the addition of water to feed in the trough (at a meal-to-water ratio of 1:1 or 1:2) can save some 5 per cent in feed usage and decrease time to slaughter by 10 days when compared with dry feeding. Feed provided wet must be totally consumed each day or it will spoil.

Meals or pellets

Pelleted feed has consistently been proven superior to meal by some 5 per cent in promoting live weight gain and feed conversion efficiency. However, such improvements in performance have to be related to the cost of the pellets. Moreover, pelleting equipment is not always available in the tropics.

Separate feeding

Small- and medium-scale operations have had some success with separate feeding of the basal feed and the protein concentrate. Systems in Madagascar, for example, supply fresh cassava roots, dug twice a week and fed directly to the pigs to appetite. Protein concentrate is fed daily according to a fixed rate (500 g for weaners up to 50 kg live weight, 750 g for finishers above 50 kg). Such 'separate feeding' systems maximise the added value from growing feed crops on the same farm. Although they may be sub-optimal in terms of nutrient balance, they optimise the economic efficiency of production.

Phase feeding

In phase feeding systems, the nutrient composition of the diet is changed (more bulk/fibre is added) according to the pig's changing requirements and increasing appetite as it grows (reduced nutrient density). Phase feeding systems are in use throughout the world, and range from simple two phase grower–finisher feeding to frequent changes in diet throughout the growing period. They improve the efficiency of feed utilisation, but demand a relatively high level of management precision. Phase feeding is particularly suited to liquid feeding and computer controlled systems.

Sex

If pigs are reared intensively and achieve fast growth rates, there is little justification for castration of male pigs (see Chapter 2). It is therefore recommended that, except for small units where it is impractical, the sexes should be segregated for fattening purposes. This allows for different feed requirements and also prevents the early-maturing boars from continually riding the gilts. However, on more extensive systems of production where males grow more slowly it may be necessary to castrate males to avoid the occurrence of boar taint in the carcass at slaughter.

Vices

Vices mainly occur in pigs that are kept in intensive confined conditions and are generally considered to be a response to boredom. Tail biting is the most common vice, but other forms of cannibalism involve biting of ears and vulvas. Once blood is drawn, very serious wounds can develop which often lead to secondary infections. Many factors can trigger tail biting, but in hot conditions, over-stocking, lack of bedding, poor ventilation and lack of salt appear to be the most important. Providing a small amount of bedding will invariably prevent tail biting. If bedding cannot be provided, some producers hang a chain from the roof and the pigs can play with this to alleviate boredom.

Killing-out percentage

The killing-out percentage is the ratio between carcass weight and live weight. It can have a significant effect on profitability. Pigs fed bulkier feeds of lower nutrient and energy density will develop a large gut capacity in proportion to body size and therefore have a markedly lower killing-out percentage. The genetics of the pig will also affect killing-out percentage, with meatier breeds tending towards a higher value.

Keeping records

Records are essential for monitoring both technical and economic efficiency of a pig enterprise. They should cover all aspects so that producers can evaluate boar performance, sow productivity, weaner growth and grower/finisher efficiency. Looking at the recent history of the enterprise enables each phase of production to be examined critically and weaknesses can be pinpointed. It is then possible to plan how to rectify any weaknesses.

Many different recording systems are available, and Tables 13–17 provide examples of the kind of recording sheets that can be used. The first essential is to keep an accurate service record of all sows and gilts in the piggery (Table 13). Farrowing and lactation performance details can then be kept on individual sow record cards (Table 14). All this information is available for transfer on to a sow lifetime record card (Table 16), which provides an accurate assessment of her contribution to herd performance. The same information can also be used to compile a record on individual boar performance (Table 15). As major differences in boar performance become evident, the lower-performing boars can be removed from the herd.

Some countries will publish figures for average production and production targets. These will be available from government advisory staff or other advisory agencies. Such figures allow the producer to assess his pigs' performance against that of other local producers.

Increasing numbers of producers are using pig management software programmes on personal computers. Such programmes can capture and store large amounts of data. Perhaps more importantly, they allow the producer to make more effective use of the data as a management tool, to highlight the strong and weak points in his/her operation. Averages can be produced on a day-to-day basis to answer questions such as:

- is gilt performance up to target?
- what is the level of pre-weaning mortality ?
- is the herd age profile as it should be, or are there too many old sows in the herd?
- is age at slaughter, grading and herd feed conversion efficiency on target?

Table 13 Sow service record sheet

Sow ear no.	Date weaned	Parity	Services						Date due	Date farrowed	No. born		Weaned		Weaning age (days)
			Date	Boar	Date	Boar	Date	Boar			Alive	Dead	No.	Date	

Table 14 Individual sow record card

Sow No. . . . . . . Date served . . . . Boar No.	
Due to farrow	Farrowed . . . .
Total born	Born alive . . . .
Pigs fostered off	Pigs fostered on . . . .
Date iron injection	
No. pigs weaned	Date weaned . . . .
Total litter weight weaned	
Average weaning weight	
Remarks	

Table 15 Boar performance record sheet

Boar No.		Date of birth		Breed		
Sow no.	Service date	Return to service	No. born	No. weaned	Total litter mass weaned	REMARKS

Table 16 Sow lifetime record card

Sow no.		Date of birth			Breed					
Litter	Weaning to service (days)	Service date	Actual farrowing date	Farrowing interval (days)	Live births	Still births	No. weaned	Pigs + or –	Litter weaning weight	REMARKS
1										
2										
3										
4										
5										
6										
7										
8										
9										
10										

Table 17 Herd performance report form

	Month	Month	Month	Six-month average	Suggested target
No. of sows					
No. of gilts					
No. of boars					
No. served to farrow					
No. farrowed					
Farrowing rate					
Pigs born alive					
Pigs born dead					
% born dead					
Total pigs born/sow					
Live pigs born/sow					
Piglet mortality					
Weaner mortality					
Fattener mortality					
Pigs sold					
Pigs born/sow/year					
Pigs sold/sow/year					

9 Processing and marketing

Transport

In the tropics, pigs may be transported long distances in unconventional vehicles (Fig 59). Transporting pigs to the slaughterhouse can subject them to extreme stress that results in death during transit, death in lairage at the slaughterhouse, or reduced meat quality in the carcass. Stress can be associated with handling at loading and unloading, the new surroundings, mixing with strange pigs, the physical discomfort of the journey and, most importantly under tropical conditions, high temperatures.

Fig 59 *In Vietnam, pigs are constrained in individual crates and transported to the slaughterhouse by boat*

Several measures can be taken to minimise stress:

- do not feed pigs for 12 hours before travel;
- handle the pigs quietly and gently at all times and avoid using sticks and prodders;
- avoid loading and travelling during the heat of the day;
- spray the pigs with cold water before loading and again in the truck;
- ensure that the loading ramp is properly designed with solid walls and is at the correct height for the cart, truck or trailer;
- provide a cover on the truck, good ventilation and adequate bedding; ensure the floors are not slippery and the sides of the truck are high enough to stop the pigs jumping out;
- if possible, subdivide animals into groups of 10 or fewer, and never mix pigs of different weights;
- do not stop on the way to the slaughterhouse.

Lairage

Similar considerations apply when the pigs are waiting to be slaughtered, since all the potential profits achieved during the growing period can be lost if death or damage occurs at this stage. The pigs should be penned under shade in small groups and sprayed with water. They should be fed only if there are long delays before slaughter. Pigs should be handled and driven quietly and gently at all times and supervised to prevent fighting. They should be penned as far away as possible from the sights and smells of the slaughtering process.

Slaughter

Stunning

For reasons of animal welfare, pigs should always be stunned before they are bled. Effective stunning ensures prompt and complete bleeding and also minimises the muscle contractions that cause pale soft exudative (PSE) meat (see 'Meat quality'). There are three main methods of stunning:

- *Mechanical*: A captive-bolt pistol or other implement is used to stun the animal.
- *Electrical*: A pair of tongs is used to apply an electrical charge to the pig's head. A current of 1.25 amps and 300–600 volts renders the pig unconscious within one second.
- *Gas:* Pigs are led into a tunnel containing 70–80 per cent carbon dioxide, which will cause them to lose consciousness within two seconds.

Bleeding

Straight after stunning, the animal should be suspended by its hind legs and the blood vessels of the neck completely severed to ensure thorough and complete bleeding. The blood should be collected in clean vessels.

Scalding and de-hairing

Immersing the carcass in water at 65–75°C loosens the hair, and it can then be removed by scraping. Any excess hair can be singed off by a flame. For the small-scale producer who is slaughtering on the farm, a drum of water over a fire is adequate for scalding purposes. Alternatively, when water is scarce, and if the skins are not used, de-hairing can be achieved by covering the carcass with a 5 cm deep layer of straw or dry grass and burning it. The skin can then be scraped to remove the carbonised surface and any remaining hair.

Evisceration

A long cut is made down the belly from the breast to the hams. To prevent the meat being contaminated, the entire length of the gut should be removed intact. Other internal organs can then be separated, and the gut emptied and cleaned away from the rest of the meat.

Meat hygiene

The carcass of a recently killed pig is an ideal breeding ground for bacteria. Maintaining hygienic conditions is therefore of paramount importance. Ideally, carcasses should be chilled immediately after slaughter, and the meat should remain chilled until it is cooked. Where refrigeration is not available, carcasses should be hung in a cool room, protected from flies by gauze, and then sold and eaten as soon as possible.

At the slaughterhouse, carcasses should be examined by a qualified meat inspector. He will examine the carcass and offal critically for signs of parasite infection (e.g. 'measly' pork, 'milk spot' livers, damaged lungs, etc. – see Chapter 7) and other departures from health. Meat that does not pass inspection is condemned and should be burnt. The carcasses from pigs slaughtered on the farm should also be examined critically to avoid the transmission of disease and parasites from pigs to humans.

Carcass and meat quality

Every country has its own system of assessing carcass quality, which usually involves grading and determines the price paid to the producer. There are four main considerations:

- *Conformation:* This refers to the shape of the carcass. Those that are poorly developed in the more valuable areas, such as the hams and loins, are downgraded. Conformation is mainly genetically controlled (see Chapter 2), but can be influenced by nutrition.
- *Degree of fatness:* The amount of fat in the carcass is normally estimated by measuring the amount of subcutaneous fat cover at a set position. In Zimbabwe, for example, this is carried out at the 'K' position, which is 7.5 cm from the mid-line in line with the last rib (Fig 60). In the UK the 'P2' position is used which is 6.5 cm from the mid-line. The fat measurement that qualifies a carcass for top grade will then be decided according to market demand.
- *Lean content of the carcass:* A major fault of carcass grading systems based on measuring the fat cover is that no account is taken of the amount of lean meat in the carcass. In fact, the meatier pigs in a group will often have the heaviest fat cover and will be downgraded. Recently, equipment has been developed that can measure muscle depth in the carcass and estimate the total lean content. As a result, systems that grade carcasses according to the degree of lean meat in the carcass are now in operation in some developing countries.
- *Fat quality:* Carcasses exhibiting a soft and oily, rather than hard, fat cover will tend to be downgraded because they cannot be used in

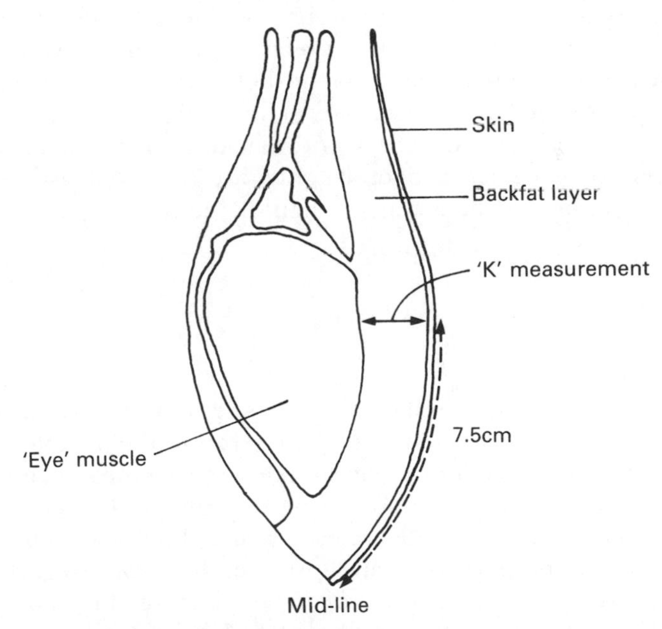

Fig 60 *The 'K' back-fat measurement used for grading in Zimbabwe, measured on a cross-section of the half carcass in line with the tip of the last rib*

higher-priced fresh or cured meat products. The cause of soft fat can be genetic, but is more likely to be due to a high proportion of unsaturated fats in the diet. Uneven growth rates causing frequent mobilisation and re-deposition of fat can also be contributing factors.

Meat quality (as distinct from carcass quality) is a measure of the desirability of the meat for the consumer. It is related to colour, texture, flavour and odour. Pig meat is characteristically paler and more tender than the meat of ruminants, but extremes do occur. Very pale, watery meat is associated with the condition known as pale, soft, exudative (PSE) muscle, which occurs in pigs that are homozygous for the 'halothane' gene (see Chapter 4), or as a consequence of poor pre-slaughter handling, or both. Such carcasses show a rapid decrease in acidity at slaughter, which reduces the water-holding capacity of the lean tissue, leading to 'drip-loss' from the meat and excessive paleness. At the other extreme is a condition known as dry, firm, dark (DFD) muscle, which can be attributed directly to pre-slaughter stress. As the age at which pigs are slaughtered increases there is a corresponding reduction in the paleness and tenderness of the meat produced.

Meat texture can be influenced by the amount of intramuscular or 'marbling' fat deposited. This can increase from 1 per cent in lean pigs to about 3 per cent in fat pigs, and can improve the succulence and eating quality of the meat. The Duroc breed tends to have a higher percentage of intramuscular fat than the white breeds. Meat produced from this breed and its crosses will thus have a relatively high meat quality.

Pig meat can pick up flavours from the diet, particularly if high levels of fish meals or rancid fats are fed. However, the main concern is boar taint, which can occur in the meat from entire males. The problem does not occur if entire males destined for slaughter are grown out fast enough to allow them to be slaughtered at less than six months of age.

Marketing

Uses of pig meat

The final phase in pig production is the sale and disposal of the end product. The pig is extremely versatile in terms of the number of products that can be derived from pig meat. Fresh meat is the most important product in the tropics, since processing facilities are limited. Pigs destined for the fresh meat trade are usually slaughtered at younger ages and lower weights (porkers) than those used for processing. If sold in the commercial market, the carcass can be butchered into a number of wholesale cuts (Fig 61), which can be cooked and eaten in a number of different ways.

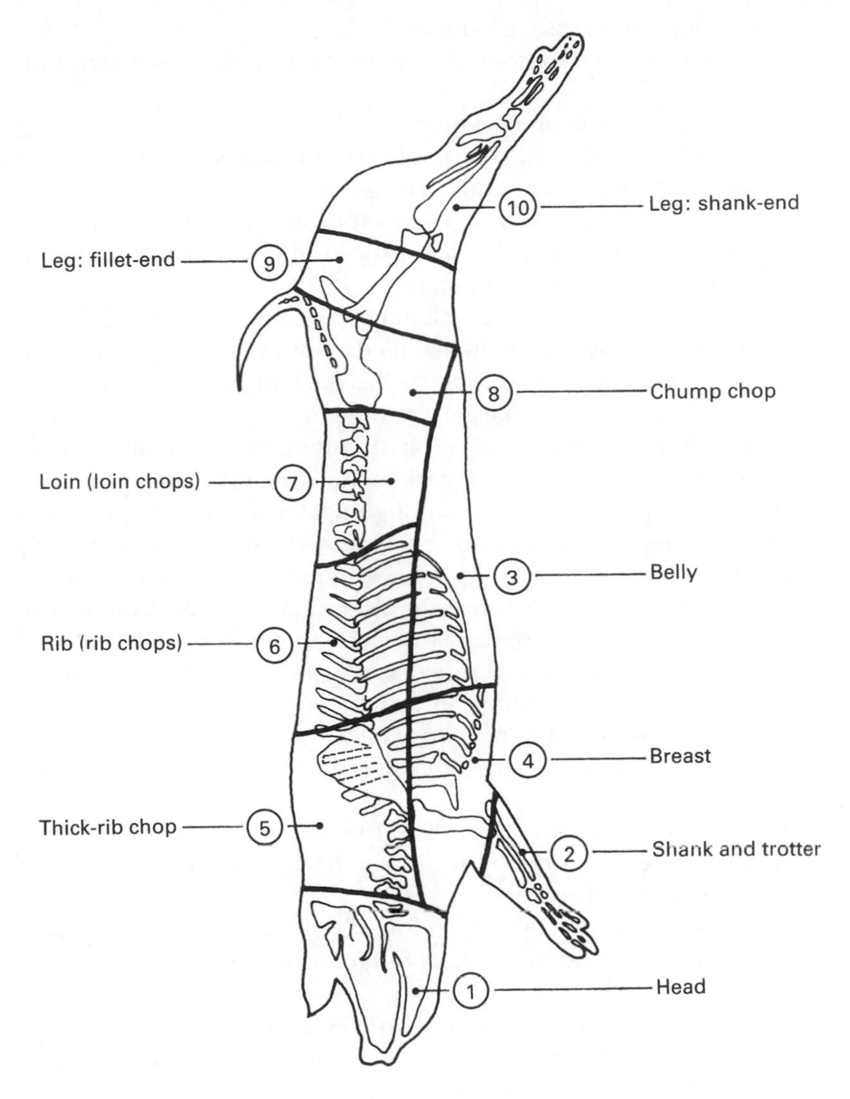

Fig 61 *Wholesale meat cuts from a pork carcass*

Cured products include various bacons and hams that are cured in brine and sometimes flavoured by hanging in smoke. Both processes increase the shelf life of the product compared with fresh meat. Bacon and hams are derived from the 'baconer' category of pig, which is heavier than the porker. Other processed products include various types of sausages, pies, luncheon meats, hamburgers and meat pastes. These tend to be produced from the lower-value portions of the porker and baconer carcasses, or from heavy hogs, mature sows and boars.

Additional products of pig carcasses:

- lard (pig fat) is sold for food, feed or as an ingredient in soap production;
- pigskin makes a valuable leather;
- bristles are used to make shaving or paint brushes;
- intestines are used for sausage casings;
- offals, especially the liver, are valued for human consumption;
- blood is collected separately and processed into sausages and other delicacies for human consumption;
- slaughterhouse by-products, including bones, blood and inedible meat tissue, are converted into animal feeds; however, owing to concerns of disease transmission, this practice is declining;
- hoofs are used for to make gelatin and glue.

The specific advantages of pig meat are often not fully realised by the consumer. The main benefits of pork include the following:

- *Calorific value and cholesterol*: Although pork contains a high concentration of nutrients, lean pork is relatively low in calories and, of the major meats, only chicken has a lower calorific value (see Table 18). Pork has a relatively low cholesterol level at 0.009 per cent of lean meat. These features appeal particularly to consumers in the developed world, but there is an increasing trend for consumers in the developing world, especially in urban and peri-urban areas, to become more aware of the need to reduce their dietary fat.
- *Digestibility*: Pork is 98 per cent digestible by the human digestive system.
- *Protein quality*: Pork contains all eight essential amino acids (which cannot be manufactured by the human body) and is therefore a high quality, complete protein.
- *'Satiety'*: Lean pig meat has been found to have a high 'satiety' value, i.e. it leaves the consumer feeling full for long periods and helps prevent overeating.
- *Minerals*: Pork contains relatively large amounts of iron, essential for prevention of anaemia, and zinc, which hastens wound healing, bone growth and tissue development.
- *Vitamins*: The levels of the B vitamins are particularly high in pork and it is a leading dietary source of thiamine.

Systems of marketing

Private sales are the most common marketing method in the tropics among small-scale producers. One, or a number of pigs are sold to local consumers, other producers, butchers or middlemen. The pigs are sold live and the price is generally subject to negotiation. This system has the advantage of being the simplest, but in rural areas individuals who are

Table 18 Comparison of the calorific contents of the main sources of meat (based on an average portion of 85 g of lean meat)

	No. of calories per 85 g of lean[1]
Beef	228
Lamb	221
Pig's liver	220
Pork	206
Chicken	144

[1] To obtain Fig in Kj, multiply by 42

not aware of current prices can be taken advantage of by speculators and dealers. Marketing co-operatives have been formed in some rural areas to overcome this problem and ensure members achieve adequate prices. Small-scale producers may also take advantage of public sales, which involve taking the pigs to a central marketplace where they are sold by auction on a live basis to the highest bidder.

Direct selling to an abattoir or butchery is more applicable to the larger-scale producer. The big disadvantage of direct sales is the effect of the 'pig cycle'. This is the notorious fluctuation in price that occurs in most countries as a result of fluctuations in supply. When pigs are in short supply, prices rise. This, in turn, stimulates increased production, more pigs become available, and, consequently, prices fall. As it takes about a year for a producer to react to price changes, the cycle will occur every 12 to 18 months, leading to a lack of stability with producers going in and out of pig production. Larger-scale producers are advised to pursue contract sales. These involve entering into a contract with an abattoir to supply a certain number of pigs over a certain period at a set price, and largely protect the producer from the effects of the pig cycle. In turn, this allows him to plan his production output over a longer period.

New marketing opportunities

Traditionally, pig meat has provided a wide variety of marketable products (see 'Uses of pig meat' above) and the opportunity must be taken to exploit this to the maximum in the developing market situation. In this respect, there is a need to tailor pig production towards specific end products and markets.

Some opportunities include the following:

- Overall promotion of the qualities of pig meat, emphasising the nutritional and health benefits to the consumer.
- Providing the quality expectations of visiting consumers from the developed world (tourists, businessmen, expatriates). This requires low

levels of subcutaneous fat, firm white fat, good meat texture and succulence.

- Increasing the variety of products by more processing, especially of the cheaper cuts.
- Providing small portions for low-income urban consumers.
- Exploring niche markets, e.g. the trend towards 'organically produced meat'. Tropical producers are well placed to take advantage of this market due to more extensive, 'welfare-friendly' systems of production, low use of drugs and chemicals, etc.

A final word

During the past two decades, there has been a tremendous increase in our level of knowledge of the physiology and nutrition of the pig. This has allowed us to design systems of pig production that allow pigs to be raised in environments that would previously have been considered too hostile. Tropical developing regions of the world are prime examples of such environments, where high levels of temperature, humidity, solar radiation and disease are common. Introducing relatively simple modifications of the environment now allows producers in the developing world to rear superior genotypes of pig and to produce quality pig meat in a cost-effective manner.

Appendix A

A guide to the nutrient requirements of pigs (with consideration of the likely availability of feed resources in tropical regions)

Concentration in diet ('as fed' basis, i.e. moisture content ~10%)	Growing pigs bodyweight range (kg)				Breeding pigs		
	5–10	10–20	20–65	65–120	Pregnant sows	Lactating sows	Boars
Digestible energy (MJ/kg)	13.5	13	12.5	12.5	12.5	12.5	12.5
Crude protein (g/kg)	220	190	170	140	120	140	140
Lysine (g/kg)	13	10	9	7	5	7	7
Methionine + cystine (g/kg)	7	5	4.5	4	3.5	4	4
Tryptophan (g/kg)	2	1.9	1.7	1.4	0.8	1.4	1.4
Calcium (g/kg)	7	7	7	5.5	7.5	7.5	5.5
Phosphorus (g/kg)	6	6	5.5	4.5	5	5	4.5
Sodium chloride (g/kg)	5	5	5	5	5	5	5
Magnesium (g/kg)	0.4	0.4	0.4	0.4	0.4	0.4	0.4
Manganese (mg/kg)	20	20	20	20	20	20	20
Zinc (mg/kg)	100	100	100	100	50	50	100
Iron (mg/kg)	100	100	150	150	80	80	150
Copper (mg/kg)	10	7	7	7	7	7	7
Iodine (mg/kg)	0.14	0.14	0.16	0.16	0.5	0.5	0.2
Selenium (mg/kg)	0.15	0.15	0.15	0.15	0.15	0.15	0.15
Vitamin A (i.u./kg)	6000	6000	4000	4000	6000	4000	4000
Vitamin D (i.u/kg)	2500	2000	1500	1500	2000	1500	1500
Vitamin E (mg/kg)	10	10	10	10	20	20	10
Biotin (mg/kg)	0.3	0.3	0.3	0.3	0.3	0.3	0.3
Riboflavin (mg/kg)	5	4	4	3	4	4	3
Thiamin (mg/kg)	1.3	1.1	1.1	1.1	1.1	1.1	1.1
Vitamin B12 (mg/kg)	18	18	15	15	15	15	15
Pantothenic acid (mg/kg)	15	15	10	10	15	15	10
Choline (mg/kg)	500	500	500	500	550	550	500

Appendix B

The nutrient composition of the major feedstuffs available for pig feeding in the tropics ('as-fed' basis)

	Macro-nutrients						
	Dry matter	DE	Crude protein	Fibre	Lysine	Methio-nine + cystine	Trypto-phan
	%	MJ/kg	%	%	%	%	%
1. Energy sources							
(a) Cereals and by-products							
Maize	89	13.9	8.9	2.9	0.22	0.30	0.09
Maize germ meal	92	11.5	20.0	10.9	0.90	0.85	0.25
Maize bran	89	10.5	10.1	9.0	0.27	0.36	–
Millets	90	12.7	11.5	6.5	0.23	0.48	0.17
Rice	89	10.4	7.3	10.0	0.24	0.22	0.12
Rice bran	89	9.7	13.5	13.0	0.50	0.27	0.10
Rice polishings	90	13.2	11.0	4.0	0.49	0.28	0.14
Sorghum	90	14.2	11.0	2.0	0.27	0.30	0.09
Wheat	89	15.1	10.8	2.8	0.30	0.34	0.12
Wheat bran	89	10.2	14.8	10.0	0.60	0.50	0.30
(b) Other energy sources							
Bananas	22	3.1	1.3	0.8	0.06	0.03	–
Cassava tubers	63	10.5	1.7	5.5	–	–	–
Cassava meal	87	14.6	2.4	7.6	–	–	–
Molasses	74	10.3	2.9	–	–	–	–
Potatoes	23	3.6	2.2	0.7	0.11	0.05	0.02
Sweet potatoes	32	5.0	1.4	0.5	0.06	0.04	0.03
Yams	25	2.7	2.5	0.7	0.12	0.08	–

	Macro-nutrients						
	Dry matter	DE	Crude protein	Fibre	Lysine	Methio-nine + cystine	Trypto-phan
	%	MJ/kg	%	%	%	%	%
2. Protein sources							
Blood meal	89	8.5	80.0	1.0	5.3	2.40	1.00
Chickpeas	89	10.8	20.1	9.0	1.4	0.45	0.07
Coconut oil meal	93	11.0	22.0	12.0	0.5	0.53	0.20
Cottonseed meal	91	10.8	41.0	11.0	1.6	1.60	0.70
Feather meal	93	10.0	85.0	1.5	1.1	3.60	0.40
Fish meal	91	10.2	62.0	1.0	5.3	2.40	0.69
Groundnut meal	92	10.7	48.0	6.8	2.3	1.20	0.39
Kidney beans	89	13.6	25.7	8.2	1.7	0.40	0.70
Linseed oil meal	88	8.8	33.0	9.5	1.1	1.10	0.48
Lucerne meal	92	8.9	20.0	27.0	0.9	0.56	0.46
Meat meal[1]	93	11.2	55.0	2.5	3.0	1.43	0.35
Meat-and-bone meal[1]	92	10.8	45.0	2.5	2.2	0.95	0.18
Palm kernel meal	90	10.5	17.0	15.0	0.6	0.78	0.17
Pigeon peas	89	11.9	21.0	7.0	1.6	0.31	0.04
Safflower meal	91	12.4	28.5	30.6	0.7	0.68	0.26
Sesame seed meal	94	11.3	42.0	6.5	1.3	1.50	0.82
Skimmed milk (dried)	92	14.8	33.0	-	2.6	1.40	0.45
Soyabean meal	89	12.5	46.0	6.0	2.9	1.50	0.55
Sunflower seed meal	93	11.5	42.0	13.0	1.7	1.10	0.50
3. Bulky feeds							
Brewing by-products (dried grains)	92	9.9	27.0	15.0	0.90	1.00	0.34
Banana leaves	20	1.4	2.9	4.5	0.13	0.96	–
Cassava leaves	28	0.8	7.0	3.2	0.43	0.19	0.10
Citrus pulp (dried)	91	8.3	6.0	12.2	0.20	0.19	0.06
Pumpkins and melons	12	1.4	1.5	1.5	–	–	–
Swill (average)	18	3.3	3.6	0.5	–	–	–
Water hyacinth	17	0.3	4.0	2.9	0.14	0.11	0.08

[1] An outbreak of Bovine Spongiform Encephalopathy (BSE) occurred in the UK during the 1980s and 1990s as a result of feeding meat-and-bone meal to cattle. No such problems have been recorded with pigs, however, stockowners should be aware of the risk and should avoid giving pig feed containing these substances to cattle.

	Minerals								
	Cal-cium %	Phos-phorus %	Sodium chloride %	Mag-nesium %	Copper mg/kg	Iron mg/kg	Zinc mg/kg	Man-ganese mg/kg	Iodine mg/kg
1. Energy sources									
(a) Cereals and by-products									
Maize	0.01	0.25	0.08	0.12	5.80	35	23.0	5.0	4.5
Maize germ meal	0.04	0.58	0.25	0.30	–	–	18.0	–	–
Maize bran	0.03	0.23	0.05	0.21	–	–	–	0.7	–
Millets	0.05	0.30	0.05	–	–	–	15.0	35.0	–
Rice	0.07	0.31	0.04	0.12	-	20	4.0	12.0	4.0
Rice bran	0.10	1.70	0.35	0.95	13.00	190	30.0	138.0	–
Rice polishings	0.04	1.40	0.18	0.65	–	160	–	–	–
Sorghum	0.04	0.29	0.13	0.13	13.00	53	15.0	15.0	0.02
Wheat	0.05	0.30	0.13	0.15	0.04	50	15.0	39.0	7.8
Wheat bran	0.14	1.17	0.12	0.55	12.30	200	70.0	116.0	12.0
(b) Other energy sources									
Bananas	0.01	0.03	0.19	0.40	–	–	–	–	–
Cassava tubers	0.11	0.06	0.01	0.08	–	–	–	–	–
Cassava meal	0.15	0.08	0.02	0.13	–	–	–	–	–
Molasses	0.82	0.08	-	0.35	59.60	200	–	42.2	–
Potatoes	0.01	0.05	0.08	0.03	3.70	–	–	8.8	–
Sweet potatoes	0.04	0.05	0.04	0.05	–	–	–	–	–
Yams	–	0.04	–	–	–	–	–	–	–
2. Protein sources									
Blood meal	0.28	0.22	1.50	0.22	9.90	3000	–	5.3	–
Chickpeas	0.26	0.32	0.10	–	–	–	30.0	–	–
Coconut oil meal	0.17	0.60	0.10	0.30	–	–	–	–	–
Cottonseed meal	0.18	1.20	0.05	0.54	17.60	100	–	23.1	0.1
Feather meal	0.20	0.70	0.71	0.20	–	–	–	21.0	–
Fish meal	5.00	2.70	1.50	0.20	1.10	270	68.0	40.0	20.0
Groundnut meal	0.16	0.52	0.12	0.22	–	20	–	18.0	30.0
Kidney beans	0.14	0.54	0.80	0.13	4.10	70	42.0	8.4	–
Linseed oil meal	0.35	0.75	–	0.60	25.70	0.03	–	37.6	–
Lucerne meal	1.50	0.27	0.50	0.27	0.50	212	35.0	86.0	10.0
Meat meal	7.60	4.00	3.50	0.27	8.00	410	–	18.0	–
Meat-and-bone meal	11.00	5.90	1.80	1.13	1.30	500	98.0	10.0	12.0
Palm kernel meal	0.40	0.50	0.06	0.35	–	–	–	–	–
Pigeon peas	0.24	0.30	0.10	–	–	50	30.0	–	–

	Minerals								
	Cal-cium %	Phos-phorus %	Sodium chloride %	Mag-nesium %	Copper mg/kg	Iron mg/kg	Zinc mg/kg	Man-ganese mg/kg	Iodine mg/kg
Safflower meal	0.40	1.10	–	0.37	10.80	560.0	44.0	19.8	–
Sesame seed meal	2.00	1.30	0.10	0.80	–	–	100.0	47.9	–
Skimmed milk (dried)	1.30	1.00	0.94	0.11	11.50	50.0	40.0	2.0	–
Soyabean meal	0.25	0.60	0.34	0.28	0.13	150.0	28.5	30.0	70.0
Sunflower seed meal	0.41	0.95	–	–	–	30.0	–	23.0	–
3. Bulky feeds									
Brewing by-products (dried grains)	0.27	0.50	0.26	0.14	21.30	2.5	98.0	37.6	–
Banana leaves	0.11	0.07	–	–	–	–	–	–	–
Cassava leaves	0.14	0.03	0.06	–	–	–	–	–	–
Citrus pulp (dried)	1.40	0.10	0.16	0.16	5.70	1.6	14.5	6.8	–
Pumpkins and melons	–	–	–	–	–	–	–	–	–
Swill (average)	–	–	–	–	–	–	–	–	–
Water hyacinth	–	0.16	–	–	–	–	–	–	–

	Vitamins							
	A i.u./kg	E mg/kg	Biotin mg/kg	Ribo-flavin mg/kg	Thiamin mg/kg	B12 µg/kg	Pantoth-enic acid mg/kg	Choline mg/kg
1. Energy sources								
(a) Cereals and by-products								
Maize	400	22.0	0.23	1.3	4.5	0.21	3.9	400
Maize germ meal	–	91.0	–	3.0	20.0	–	12.0	400
Maize bran	–	–	0.10	1.5	4.4	–	5.3	–
Millets	510	–	–	1.6	6.6	–	7.4	789
Rice	–	–	0.10	1.2	2.4	–	8.6	900
Rice bran	–	1.0	4.20	3.3	5.0	–	14.0	1135
Rice polishings	–	4.0	0.61	11.0	7.3	–	53.0	1237
Sorghum	740	24.0	0.28	1.0	3.6	–	9.0	700
Wheat	250	34.0	0.11	1.0	4.6	1.20	13.0	660
Wheat bran	–	99.0	–	3.1	7.9	–	29.0	1000

	Vitamins							
	A	E	Biotin	Ribo-flavin	Thiamin	B12	Pantoth-enic acid	Choline
	i.u./kg	mg/kg	mg/kg	mg/kg	mg/kg	µg/kg	mg/kg	mg/kg
(b) Other energy sources								
Bananas	55	–	–	0.50	0.6	–	–	–
Cassava tubers	–	–	–	0.14	–	–	–	–
Cassava meal	–	–	–	0.20	–	–	–	–
Molasses	–	–	–	3.30	0.9	-	38.3	875
Potatoes	–	–	–	0.20	1.5	–	6.4	–
Sweet potatoes	–	–	–	–	–	–	–	–
Yams	131	–	–	–	–	–	–	–
2. *Protein sources*								
Blood meal	–	–	0.10	1.50	0.4	42.00	1.2	1000
Chickpeas	–	–	0.18	0.40	0.5	–	10.0	642
Coconut oil meal	–	4.0	–	19.8	0.9	–	4.4	–
Cottonseed meal	300	24.0	1.10	5.00	5.3	–	10.3	2300
Feather meal	–	–	0.04	2.10	0.1	0.60	10.0	891
Fish meal	–	24.0	0.12	6.20	0.9	170.00	8.6	2800
Groundnut meal	300	–	0.40	10.40	0.6	–	48.0	1700
Kidney beans	–	1.0	0.09	1.60	5.5	–	3.0	1670
Linseed oil meal	–	–	–	2.90	9.5	–	–	1225
Lucerne meal	17800	240.0	0.28	13.00	3.0	5.00	26.0	800
Meat meal	–	–	0.09	5.00	0.2	55.00	5.0	2200
Meat-and-bone meal	–	–	0.10	46.00	0.2	120.00	3.0	1500
Palm kernel meal	69000	–	–	0.50	1.4	–	–	–
Pigeon peas	–	–	0.18	0.80	1.8	–	4.6	642
Safflower meal	550	0.9	1.56	11.30	2.8	–	43.8	2247
Sesame seed meal	–	–	0.34	3.60	2.8	–	6.0	1690
Skimmed milk (dried)	24000	9.1	0.33	22.00	3.5	0.01	33.0	1250
Soyabean meal	340	20.0	0.32	26.00	50.0	2.10	14.1	250
Sunflower seed meal	–	11.0	1.45	3.10	–	–	10.0	2894
3. *Bulky feeds*								
Brewing by-products (dried grains)	–	25.0	0.96	1.50	0.7	–	8.6	1587
Banana leaves	15000	–	–	0.10	0.4	–	–	–
Cassava leaves	–	–	–	0.60	0.4	–	–	–
Citrus pulp (dried)	–	–	–	–	2.4	1.50	13.0	845
Pumpkins and melons	–	–	–	–	–	–	–	–
Swill (average)	–	–	–	–	–	–	–	–
Water hyacinth	–	–	–	–	–	–	–	–

Glossary

Abortion premature expulsion of the foetus before it is able to survive
Acaricide chemical that kills ticks and mites
Ad lib (short for *ad libitum*) unrestricted feeding to appetite
Agalactia lack of milk or milk letdown in the sow
Ambient temperature the temperature in the surrounding environment
Amino acids the basic units of protein
Anaemia deficiency of haemoglobin in the blood due to blood loss, lack of dietary iron or poor iron absorption
Anoestrus failing to come into heat
Anthelmintic chemical that kills intestinal worms and parasites
Antibiotic drug synthesised by microorganisms that can inhibit the growth of other microorganisms
Antibody substance produced in the body in response to invasion by a foreign organism
Antioxidant substance that prevents or reduces the oxidation of other substances
Anti-nutritional factor factor that interferes with normal processes of digestion and absorption
Artificial insemination (AI) the collection of semen from a boar and the introduction of the semen at a later stage into a sow or gilt by means of a catheter
Bacon pig meat that has been cured in brine, with or without smoking
By-product product which is additional to the main product(s) for which an animal or process has been developed (now often called a co-product)
Caecum part of the hindgut that contains microbes able to digest some fibre in the pig
Castrate to remove the testes from a male pig
Colostrum the first milk secreted after a sow gives birth
Concentrate feed containing a high concentration of nutrients, low in fibre and highly digestible
Conformation the physical form and shape of an animal or carcass

Creep feeding making supplementary feed available to suckling piglets but not to the sow

Criss-cross breeding system when two breeds are crossed, a criss-cross system involves crossing back to each of the parent breeds alternately each subsequent generation

Cyanogenetic glucoside substance that can be broken down to produce hydrogen cyanide (prussic acid)

Dermatitis irritation and/or inflammation of the skin

Diarrhoea the frequent passage of loose, fluid faeces

Digesta products of digestion

Dressing per cent the weight of the carcass expressed as a percentage of live weight at slaughter

Encyst to enclose in a cyst

Endemic disease disease regularly found in a specified area or country

Enteric relating to the intestine

Entire male male pig with testes intact (cf. castrate)

Environment the sum of conditions affecting the well being of the animal

Enzyme biological catalyst produced by living cells

Evisceration removal of the intestines and internal organs

Farrowing the act of giving birth to piglets by the sow

Feed conversion efficiency the amount of feed required for an animal to make a unit gain in weight

Gene unit of hereditary material in the chromosome

Genotype the genetic constitution of an individual

Gestation the period from mating to birth of the young; also known as pregnancy

Gilt young female pig up to the time she produces her first litter

Growth rate the live weight gain of an animal per unit time

Haemorrhage profuse bleeding due to rupture of blood vessels

Heterozygous in simple Mendelian inheritance, when a character is controlled by a single pair of genes, the heterozygous condition is when both a dominant and recessive gene are present

Homeotherm (sometimes called endotherm) a warm-blooded animal

Homozygous in simple Mendelian inheritance, when a character is controlled by a single pair of genes, the homozygous condition is when both genes are either dominant (homozygous dominant) or recessive (homozygous recessive)

Hormone chemical 'messenger' secreted by an organ of the body and conveyed in the blood to regulate the function of tissue or organs elsewhere

Hybrid the result of a cross between purebred animals

Hyperthermia an increase in the deep body temperature of an animal

Hypothermia a decrease in the deep body temperature of an animal
Ill-thrift failure to thrive normally; poor condition
Immunity the ability to resist and overcome infection
Insulin hormone involved in regulating carbohydrate metabolism
In utero inside the uterus
Lacrimation the secretion of tears from the eyes
Lactation the period when the sow is producing milk for her young
Lard type a breed of pig developed to produce large amounts of fat in the carcass
Libido sexual drive and energy
Litter the young produced by a sow at a single birth
Maintenance level when the requirements of the animal for nutrients for vital processes when at rest are just met, and there is no gain or loss of nutrients
Mastication the action of chewing
Mycoplasma small organisms intermediate in size between viruses and bacteria and associated with various infections
Notifiable disease disease which must by law be reported to the government veterinary authorities
Oedema accumulation of fluid in the body, giving rise to swelling
Oestrus the period during which the female will stand and permit the male to mate
Omnivore has the ability to eat any type of food or feed e.g. humans and pigs
Osteomalacia infectious mineralisation of the mature skeleton with softening and bone pain, commonly caused by deficiency in vitamin D
Parity number of litters of piglets bred by a sow
Peptidase enzyme that breaks down proteins into amino acids
Per capita for each person
Phenotypic the physical appearance of the animal
Placentation formation of the placenta
Pork fresh pig meat, normally derived from pigs slaughtered at less than 70 kg live weight
Prophylactic course of action taken to prevent disease
Puberty defines the state when the young animal first becomes capable of breeding
Rancid applies to fats and oils that have become oxidised giving a bad smell and taste
Rickets disorder of calcium and phosphorus metabolism associated with deficiency of vitamin D; causes a softening and bending of the weight-bearing bones
Ruminant animal possessing a large first stomach known as a rumen, which is capable of digesting large quantities of fibrous material

Scour persistent watery diarrhoea
Spirochaete spiral-shaped bacterium
Subcutaneous underneath the skin
Systemic affecting the entire body
Terminal sire boar used solely to produce progeny for slaughter
Therapeutic course of action concerned with treating and curing a disease
Unthriftiness failure to thrive normally – same as ill-thrift
Vaccine suspension of dead or weakened organisms, which stimulates the immune response when introduced into the body
Wallow small pond of mud or water in which pigs can roll around to keep cool
Weaning the act of separating the young pigs from their mother

Bibliography

Agricultural Research Council (ARC) 1981. *The Nutrient Requirements of Pigs.* CABI Publishing: Slough, UK.

Cole, D.J.A., Hardy, B. and Lewis, D. 1972. Nutrient density of pig diets. In: Cole, D.J.A. (ed) *Pig Production.* Pennsylvania State University Press: USA.

Chigaru, P.R.N., Maundura, L. and Holness, D. 1981. Comparative growth, food conversion efficiency and carcass composition of indigenous and Large White pigs. *Zimbabwe Journal of Agricultural Research* 19: 31–36.

DEFRA 2004. Condition scoring of pigs. Department of Environment, Food and Rural Affairs: London, UK. Available online at www.defra.gov.uk/animalh/welfare/farmed/pigs/pb3480/pigsc005.htm

English, P.R., Fowler, V.R., Baxter, S. and Smith, W.J. 1988. *The Growing and Finishing Pig: Improving Efficiency.* Farming Press: Ipswich, UK.

English, P.R., Burgess, G., Cochran, R.S. and Dunne, J. 1992. *Stockmanship: improving the care of the pig and other livestock.* Farming Press: Ipswich, UK.

Epstein, H. 1971. *Domestic Animals of China.* Holmes & Meier: New York, USA.

Epstein, H. 1971. *The Origin of the Domestic Animals of Africa.* Holmes & Meier: New York, USA.

Eusebio, J.A. 1980. *Pig Production in the Tropics.* Longman: Harlow, UK.

FAO 1999. *Quarterly Bulletins of Statistics.* Food and Agriculture Organisation of the United Nations (FAO): Rome, Italy.

Fuller, M.F. 2004. *The Encyclopaedia of Farm Animal Nutrition.* CABI Publishing: Slough, UK.

Mason, I.L. and Maule, J.P. 1960. The indigenous livestock of eastern and southern Africa. Commonwealth Agricultural Bureau. Publication No. 14.

Muirhead, M.R. and Alexander, T.J.L. 1999. *A Pocket Guide to Recognising and Treating Pig Disease.* Nottingham University Press: Nottingham, UK.

Muys, D. and Westenbrink, G. 2002. *Pig Husbandry in the Tropics.* Revised Edition. CTA/Agromisa: Wageningen, the Netherlands.

Pathiraja, N. 1986. Improvement of pig production in developing countries. 1. Exploitation of hybrid vigour (heterosis). *World Animal Review,* 60: 18–25.

Pathiraja, N. 1986. Improvement of pig production in developing countries. 2. Selection schemes. *World Animal Review,* 61: 2–10.

Perez, R. 1997. *Feeding Pigs in the Tropics.* Food and Agriculture Organisation of the United Nations (FAO): Rome, Italy.

Pond, W.G. and Maner, J.H. 1974. *Swine Production in Temperate and Tropical Environments.* W.H. Freeman and Co: San Francisco, USA.

Serres, H. and Wiseman, J. 1992. *Manual of Pig Production in the Tropics.* CABI Publishing: Slough, UK.

Walters, J.R. 1981. Peri-urban piggeries in Papua New Guinea. In: Smith, A.J. and Gunn, R.G. (eds) *Intensive Animal Production in Developing Countries.* British Society of Animal Production: Thames Ditton, UK.

Whittemore, C.T. 1987. *Elements of Pig Science.* Longman: Harlow, UK.

Whittemore, C.T. 1998. *The Science and Practice of Pig Production.* Second Edition. Blackwell Science: Oxford, UK.

Whittemore, C.T., Green, D.M. and Knap, P.W. 2001. Technical review of the energy and protein requirements of growing pigs. *Animal Science,* 73: 363–373.

Index

Note: page numbers in *italics* refer to figures, tables or illustrations